Lecture Notes of the Institute
for Computer Sciences, Social Informatics
and Telecommunications Engineering 102

Francisco Martins   Luís Lopes
Hervé Paulino (Eds.)

Sensor Systems and Software

Third International ICST Conference,
S-Cube 2012
Lisbon, Portugal, June 4-5, 2012
Revised Selected Papers

 Springer

Volume Editors

Francisco Martins
Universidade de Lisboa, Faculdade de Ciências & LaSIGE
Departamento de Informática
Faculdade de Ciências
Campo Grande, 1749-016 Lisboa, Portugal
E-mail: fmartins@di.fc.ul.pt

Luís Lopes
Universidade do Porto
CRACS/INESC-TEC & Faculdade de Ciências
Departamento de Ciência de Computadores
Faculdade de Ciências
Rua do Campo Alegre 1021, 4169-007 Porto, Portugal
E-mail: lblopes@dcc.fc.up.pt

Hervé Paulino
Universidade Nova de Lisboa
CITI & Departamento de Informática
Faculdade de Ciências e Tecnologia
2829-516 Caparica, Portugal
E-mail: herve@di.fct.unl.pt

ISSN 1867-8211
ISBN 978-3-642-32777-3
DOI 10.1007/978-3-642-32778-0

e-ISSN 1867-822X
e-ISBN 978-3-642-32778-0

Springer Heidelberg Dordrecht London New York

Library of Congress Control Number: 2012944425

CR Subject Classification (1998): C.2.1, C.2, C.3, J.2, J.3, H.2.8, C.4, K.6

Typesetting: Camera-ready by author, data conversion by Scientific Publishing Services, Chennai, India

Printed on acid-free paper

Springer is part of Springer Science+Business Media (www.springer.com)

Preface

The Third International ICST Conference on Sensor Systems and Software (S-cube 2012) was held during June 4–5, in Lisbon, Portugal. The conference focuses on the research challenges arising from the development of software systems for wireless sensor networks (WSNs). WSNs introduce innovative and interesting application scenarios that may support a large amount of different applications including environmental monitoring, disaster prevention, building automation, object tracking, nuclear reactor control, fire detection, agriculture, healthcare, and traffic monitoring. The widespread acceptance of these new services can be improved by the development of novel programming paradigms, middleware, and technologies that have the potential to radically simplify the development and deployment of WSN applications. Such improvements require contributions from many fields of expertise such as embedded systems, distributed systems, data management, system security, and machine learning. The emphasis is, thus, on layers well above the traditional MAC, routing, and transport layer protocols.

This year's technical program included four invited talks/papers, seven regular papers, and three demos. Overall we received 18 submissions corresponding to an acceptance rate of 38%. The papers make quite diverse contributions with an emphasis on: middleware, frameworks, learning from sensor data streams, stock management, e-health, and Web of Things.

There were around 25 registrants for the conference. The social program included a dinner at "A Severa," a picturesque restaurant in downtown Lisbon, where the attendees had the opportunity to listen to "fado," a music style that is characterstic of Portugal and that has recently been classified as part of the World Cultural Heritage.

Conference Organization

Steering Committee

Imrich Chlamtac   Create-Net, Italy
Sabrina Sicari   Università degli studi dell'Insubria, Italy
Stephen Hailes   University College of London, UK

Organizing Committee

Conference General Chair

Francisco Martins   University of Lisbon, Portugal

TPC Chair

Luís Lopes   University of Porto, Portugal

Local Chair

Hervé Paulino   New University of Lisbon, Portugal

Conference Coordinator

Justina Senkus   ICST

Web Chair

Ana Lúcia Rodrigues   University of Lisbon, Portugal
Rúben Quintas   University of Lisbon, Portugal

Technical Program Committee

Alberto Coen Porisini   Università degli Studi dell'Insubria, Italy
Alicia Asín Pérez   Libelium, Spain
Animesh Pathak   INRIA, France
Fausto Vieira   Instituto de Telecomunicações, Portugal
Francisco Martins   Universidade de Lisboa, Portugal
Hervé Paulino   Universidade Nova de Lisboa, Portugal
JeongGil Ko   Johns Hopkins University, USA
João Gama   Universidade do Porto, Portugal
Luís Lopes   University of Porto, Portugal
Mattia Monga   Università degli Studi di Milano, Italy

Ming Zhao	Florida International University, USA
Mirco Musolesi	University of St. Andrews, UK
Pedro Brandão	Universidade do Porto, Portugal
Pedro Pereira Rodrigues	Universidade do Porto, Portugal
Philip Morrow	University of Ulster, UK
Sabrina Sicari	Universitá degli Studi dell'Insubria, Italy
Traian Abrudan	Instituto de Telecomunicações, Portugal
Ming Zhao	Florida International University, USA

Table of Contents

Day One

Day Two

Vehicular Sensing:
Emergence of a Massive Urban Scanner

Michel Ferreira[1], Ricardo Fernandes[1], Hugo Conceição[1], Pedro Gomes[1],
Pedro M. d'Orey[1], Luís Moreira-Matias[2], João Gama[5],
Fernanda Lima[3], and Luís Damas[4]

[1] Instituto de Telecomunicações, DCC/FC, Universidade do Porto, Porto, Portugal
{michel,rjf,hc,prg,pedro.dorey}@dcc.fc.up.pt
[2] LIAAD-INESC Porto and DEI/FE, Universidade do Porto, Porto, Portugal
luis.matias@fe.up.pt
[3] Instituto Militar de Engenharia, S.E. Engenharia Cartográfica,
Praça General Tibúrcio 80, Rio de Janeiro, Brasil
fernandalins@ime.eb.br
[4] Geolink Lda., Avenida de França, 20 Sala 605, Porto, Portugal
luis@geolink.pt
[5] LIAAD - INESC Porto L.A., FEP, University of Porto, Porto, Portugal
jgama@fep.up.pt

Abstract. Vehicular sensing is emerging as a powerful mean to collect information using the variety of sensors that equip modern vehicles. These sensors range from simple speedometers to complex video capturing systems capable of performing image recognition. The advent of connected vehicles makes such information accessible nearly in real-time and creates a sensing network with a massive reach, amplified by the inherent mobility of vehicles. In this paper we discuss several applications that rely on vehicular sensing, using sensors such as the GPS receiver, windshield cameras, or specific sensors in special vehicles, such as a taximeter in taxi cabs. We further discuss connectivity issues related to the mobility and limited wireless range of an infrastructure-less network based only on vehicular nodes.

1 Introduction

Sensor networks build upon the combination of the actions of sensing and networking. Until recently, networking was absent of the vehicular environment and vehicles were not envisioned as nodes of a super-large-scale sensor network. Clearly, the sensing component has always been particularly rich in vehicles, namely in road vehicles such as cars, buses and trucks. Components such as speedometers or more technical mass air flow and oxygen sensors have been part of vehicles for decades. In addition to these myriad of sensors that are factory-installed in production vehicles, the power reserves of these vehicles are also incommensurably larger than that of the typical mote device, allowing to power not only such large set of active sensors, but also the comprehensive number of computers that equip

F. Martins, L. Lopes, and H. Paulino (Eds.): S-Cube 2012, LNICST 102, pp. 1–14, 2012.

modern vehicles. With an estimated number of 800 million vehicles in the world, the advent of vehicular networking can connect the isolated myriad of sensors in each vehicle and create an extremely powerful sensor network.

Wireless networking in vehicular environments has existed based on cellular systems for some decades. Second Generation (2G) systems allowed data communications at the maximum rate of 9.6 kbps. General Packet Radio Service (GPRS)-based systems (or 2.5G) allow for data rates up to 171 kbps and are still heavily used to implemented real-time vehicle tracking based on Global Positioning System (GPS) localization. Third Generation (3G) and Fourth Generation (4G) systems allow for much higher data rates. However, the level of interconnection of such mobile communication devices and the sensors in the vehicles is often very limited or non-existent. In professional fleets, after-market devices are commonly installed in vehicles to allow real-time monitoring of their GPS position, speed or engine mode. Professional telemetry systems designed for trucking companies are even starting to remotely evaluate several parameters that allow quantifying the performance of each driver in terms of eco-driving.

In the mass market of non-professional vehicles, the European Commission (EC) has recently approved a recommendation [6] towards the inclusion of a wireless communication device interconnected to vehicular sensors, such as the deceleration sensors that activate airbags. The proposed emergency Call (eCall) system can automatically dial an emergency telephone number and transmit location data in case of an accident, without any human intervention. This recommendation applies to all new car models launched after September 2011.

Wireless networking for vehicles has also been evolving through a new short to medium range communication technology operating in the 5.9 GHz band, known as Dedicated Short-Range Communication (DSRC) [18]. As these new wireless networking devices are factory-installed in the vehicles, the level of interconnection with the in-vehicle sensors can be very high, namely through the Controller Area Network (CAN) bus. New collaborative Vehicle-to-Vehicle (V2V) warning systems have been proposed, based for instance on the activation of traction sensors over slippery roads, and can create just-in-time and just-in-place virtual road signs displayed on the windshield of nearby vehicles.

In this paper we describe several applications of vehicular sensing, ranging from automatic map construction and updating, to an urban surveillance system based on windshield-installed cameras. Much of the results we present are based on experiments with a large vehicular testbed that has been deployed in the city of Porto, in Portugal. This vehicular testbed comprises 441 taxi cabs from the largest taxi company operating in the city.

The remainder of this paper is organized as follows. In the next section we describe the technological evolution of the variety of sensors that have been installed in vehicles. We then discuss the increased connectivity and sensing range enabled by the mobility of the sensor nodes. Then we describe three concrete applications through vehicular sensing: automatic road map construction and updating; camera-based search of vehicles based on automatic license plate recognition; and the prediction of taxi demand per taxi stand based on machine learning over historical and real-time streamed data of geo-localized taximeter activations. We end with some conclusions.

2   Sensors in Vehicles

Sensors have been installed in vehicles since the early days of the large-scale manufacturing of automobiles. One of the first examples is the speedometer, which measures and displays the instantaneous velocity of the vehicle. In [27] sensors are defined as devices that transform (or transduce) physical quantities such as pressure or acceleration (called measurands) into output signals (usually electrical) that serve as inputs for control systems. This definition has been extended since sensors' outputs are now used as input for a wide range of applications. Modern vehicles are equipped with sensors that allow collecting information on the vehicle state, its surrounding environment and, more recently, on the driver condition.

   In the following, a short survey of the main in-vehicle sensing devices relevant for networked vehicular sensing is given. These devices are categorized according to the measuring principle (e.g. radar)or the primary usage application (e.g. automotive). The most common devices are *Automotive* sensors, which measure motion and pressure characteristics. For instance, the Anti-lock Braking System (ABS) is able to detect rapid decelerations, traction control allows to infer varying road surface conditions and belt pressure sensors can be used to determine vehicle occupancy. *Environmental* sensing devices allow to monitor the external surroundings of the vehicle. Light and rain sensors detectors are good examples of this class. In [16] and in the URBISNET [1] project the authors propose the deployment of sensors (measuring the concentrations of gases) in the public transportation system for monitoring urban pollution levels. In [28] Rodrigues et al. present a system that monitors the heart wave and data from other in-vehicle sources to better understand driver behavior. *Radar* use radio waves to detect a variety of objects (e.g. pedestrians or other vehicles). Many modern cars are equipped with short range radars (e.g. for parking assistance) and with long range radars (e.g. for anti-collision). *Cameras* are used as sensors for many applications. Image processing and pattern recognition techniques allow the development of a variety of applications, such as, lane departure warning, traffic sign/light recognition and object recognition. Lastly, *Communication* devices can be used to add geographical information and communicate measurement of other sensors but also as sensors for vehicular sensing. For example, GPS devices associated with communication capabilities can be used to perform real time monitoring of traffic conditions (e.g. [22]) or of a fleet of vehicles. Bluetooth devices available at most cars (e.g. hands free) can be used to capture traffic flow and to predict travel times and origin/destination matrix estimation [2].

   With the evolution of the automotive industry, the number of in-vehicle sensors is increasing rapidly. The interconnection of these sensors is a complex task due to the distributed wired infrastructure and stringent latency times. The issue of intra-vehicle communication has been considered in [25,12]. The autonomous vehicle may be the next big thing in the automotive industry.

Several companies like Volkswagen, Mercedes, General Motors and Google are working on their own version of an autonomous vehicle [1]. These new vehicles incorporate new sensing devices, such as, laser-beam range finders and position estimators.

3   The Increased Range of Mobile Sensors

An interesting characteristic of vehicular sensing is that the sensors are highly mobile. While this mobility can create some difficulties regarding the wireless networking aspect of inter-vehicle communication, it can greatly amplify the sensing range of each sensor. In [8] and [9] we have studied the delay-tolerant connectivity of a vehicular sensor network in an urban scenario. We have defined this delay-tolerant connectivity as a function of the time interval considered. Instead of considering just the instantaneous connectivity of the vehicular sensors based on short-ranged inter-vehicle communication, we considered the achieved connectivity after allowing the vehicles to move for a period of time. Computing the transitive connectivity over such an interval of time has to be done considering only inter-node paths that maintain a certain time order between the travelled wireless links. Following this, the transitive closure $\mathcal{L}'*$ of the resultant network is as follows:

$$\forall u_i, u_j \in \mathcal{U}, t' \in [0...t],$$
$$\{u_i, u_j, t'\} \in \mathcal{L}'^* \iff \begin{cases} u_j \in \mathcal{N}(u_i)_{t'} \\ \exists u_k \in \mathcal{U}, \{u_i, u_k, t'\} \in \mathcal{L}' \wedge \{u_k, u_j, t''\} \in \mathcal{L}'^* \wedge t' \geq t'' \end{cases} \tag{1}$$

where $\mathcal{U}$ is the set of nodes in the network, $\mathcal{N}(u_i)_{t'}$ is the set of one-hop neighbors of node u_i at time t', and $\mathcal{L}'$ is the set of links.

Using the traces generated by Development of Intervehicular Reliable Telematics (DIVERT) traffic simulator [7], we were able to understand the evolution of the connectivity in time in a large scale scenario, which was, in this case, the city of Porto, Portugal. This evolution is evident in Fig. 3. For more detailed results, we refer the reader to [8] and [9].

(a) Instant connectivity at time t.   (b) Transitive connectivity $\mathcal{L}'^*$ at $t' + 30s$.

Fig. 1. Evolution of the connectivity in time

---

[1] Recently, Google presented their driverless Toyota Prius to the media [15].

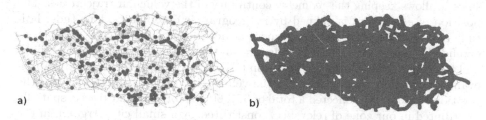

Fig. 2. Comparison of the instantaneous vs. mobile (1 hour interval) sensing range on a vehicular sensor network in the city of Porto, Portugal

In addition to the increased (delay-tolerant) connectivity of a vehicular network over a period of time, the sensing range of each sensor is also clearly amplified. In Fig. 2 we show a snapshot of the sensed area of a set of vehicular sensors over the city of Porto, in Portugal. In frame A, we have considered a sensing range of 100m per node (depicted by a red circle at each node). In frame B, we allowed nodes to move for one hour and we depict the sensed area covered by the sensors. It translated into and increase of 0.7% of the area of the city, in frame A, to 92.6% in frame B.

4   Sensing the Road Network

Since vehicles move on roads, a straightforward application of vehicular sensing is exactly related to the sensing of the road network. Based on the data collected by a large network of vehicles, it is possible to build detailed maps of highly dynamic geographic information related to this road network, such as the presence of potholes on the road [13]. Much less dynamic, but also interesting, is the information that allows mapping and updating the actual road network based on a vehicular infrastructure of remote sensing. The goal is not only the acquisition of road axle geometries, but also their characterization in terms of topological connectivity, traffic rules and speed patterns, in an accurate and permanently up-to-date manner. Several projects have been developed with the goal of making a better use of the data collected through GPS receivers [4,29]. One of the most important of these projects is the OpenStreetMap that hosts a collaborative network of GPS traces for the assisted construction of road maps. Despite the increasing research around this area, very few references relax the need of a base map in a non-assistive approach [4,11]. Most of the work presented in the literature has focused more on refinement issues and updating of existing cartography [29]. In [23] we have used a set of vehicular probes to collect data oriented to a completely automated generation of vectorial roads maps. Such automatic construction of road maps from GPS traces requires the availability of a large data set collected over the area of interest. In our implementation, we used more than 30 millions of GPS points, collected in real-time by a vehicle tracking company, using a temporal detail of one point per second. Such level of detail is particularly important for the representation of the road network, since it allows keeping the geometry continuity of the vehicular trajectories. The position of each point is defined by its geographic coordinates: longitude, latitude and altitude. To meet our purpose of accurate road map construction, we extended the protocol to also include information about the number of satellites and the Horizontal Dilution of Precision (HDOP). We also stored additional relevant information, such as speed of the vehicle, its azimuth and the time of the position reading. We collected a total of 371, 600 km of vehicular traces, spatially distributed in our zone of relevance, constituted by a small city, Arganil, in the middle of a rural area in Portugal. In Fig. 3, frame A, we depict the spatial distribution of these GPS traces.

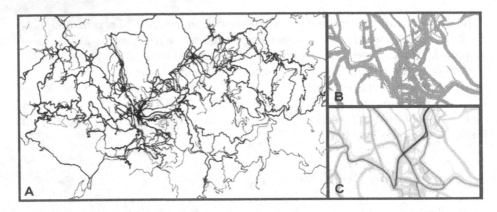

Fig. 3. Frame A depicts the set of GPS traces collected in the area of Arganil. Using such a small scale, we can easily perceive that it depicts a road network. On frame B, we highlight a small part of these road network at a much larger scale. Clearly, the noisy data reported by the GPS sensor results in a blurred description of the road network, particularly in areas where this road network is more complex. On frame C, we depict the same area as in frame B, but we use the counting of the number of GPS traces that intersects each raster cell to define the grey color intensity of the cell. Clearly, a much more defined road network starts to emerge.

Because of GPS errors, it is necessary to rely on several processes that allow the elimination of inconsistent data, aiming at obtaining higher quality input data to our algorithm. We have thus established three filters, where the first one is based on speed information. Points collected at speeds lower than 6 km/h were not considered to be sufficiently accurate for the automatic construction of road maps. As a result of this filter, we have eliminated 15,31% of the collected points. Our second filter is based on the HDOP value, which is a measure quantifying the degradation level of the horizontal positioning accuracy of the GPS (2D-based positioning). This value is mainly determined by the relative geometry of the visible satellites when the positioning reading was taken. A low HDOP value means a more accurate horizontal positioning. The filter has been configured to eliminated points with an HDOP value higher than 2.

Our third filter in the pre-processing phase of our data is based on the number of satellite used in the positioning. When more than 4 satellites are used for the positioning, the redundant satellites can serve for the detection of erroneous readings, thus increasing the accuracy of the positioning. The filter is set to eliminate points collected using a number of satellites lower than 5.

In addition to the filters mentioned above, the pre-processing phase of the data collected from the vehicles has an extra module that is responsible for the identification of large intervals of time between consecutive points of the same vehicular trace, which are either caused by obstacles to the reception of the signal broadcasted by the satellites, or by the elimination of points from previous filters. Such large intervals can erroneously affect the geometry of the road network and we thus divide vehicular traces where two consecutive points are separated by more than 7 seconds into two distinct traces. The last step in our pre-processing phase consists in the simplification of the GPS traces in order to cope with performance issues of our algorithm and minimize the amount of memory required to store the vehicular traces. Our traces have been simplified using the Douglas-Peucker algorithm [10], resulting in the elimination of 67% of the collected points. The maximum threshold distance allowed for the elimination of points through this simplification algorithm was of 1 meter.

The main goal of our algorithm is the construction of a graph representing the road network, where the roads are represented by edges and the intersections are represented by nodes (known as a network model of the road network). The algorithm is divided in five steps: rasterization, centroid generation, geometric connectivity of the centroids, topologic connectivity (node-edge topology) and turn-table construction.

4.1   Dealing with Noise from the Positioning Sensor

Until May, 2000, the real-time positioning of a point, through a navigation GPS receiver, provided a planimetric accuracy better than 100 meters. Since then, with the ending of the Selective Availability (technique used to degrade the accuracy of the positioning), such value became, on average, better than 15 meters. Even with this significant improvement, the attained accuracy is not considered to be sufficient for a valid geometric representation of the road network. As can be seen in Fig. 3, frame B, the unprocessed image that results from plotting each of the GPS points received from the vehicles, even after the filtering phase described above, creates a fuzzy image where the perception of roads is not easy. Aimed at an accurate and automatic construction of road maps, we proposed the spatial aggregation of a large set of GPS traces through a rasterization process.

The term rasterization is used in the context of the transformation of a vectorial representation into a matrix-based representation. In the work presented in [23], the rasterization process enables the transformation of the vectorial layer of GPS traces into a raster layer of 5-meter-resolution cells. For each cell, we assign a value that translates the number of GPS traces that intersect it. Using this value to vary the color intensity of each cell (depicted using different level of gray in Fig. 3, frame C), it becomes possible to easily identify the road axles. The probability of existing a road in a given cell is proportional to the value of the attribute of the cell. Similarly, cells holding a low counting value of intersecting traces represent disperse vehicular trajectories or low-travelled roads. The rasterization step thus performs an highly refined filtering of our data set, using a process based on spatial aggregation together with sampling correction. This approach becomes particularly important to the representation of small roundabouts, nearby roads and other complex parts of the road network. Furthermore, the rasterization process allows a better representation of wide roads (e.g. roads with two ways separated by a central structure), becoming possible the identification of a road axle in each of the directions.

After generating an accurate geometric map of the road network, we use again the vehicular traces to infer the connectivity layer of the identified road segments (segments between intersections. We are also able to infer the the turn-table that reflects the allowed maneuvers.

4.2   Evaluation

The evaluation phase of the results obtained was performed by comparing the geometric and topological layers of the extracted road network with those from vectorial maps provided by the map-making company InfoPortugal, S.A., which are constructed by manually processing ortho-rectified aerial images, with a resolution of $25cm^2$ per pixel. Coverage evaluation is done through three main metrics: total number of kilometers of roads generated by the algorithm in the zone of relevance; total number of kilometers of roads to which a match is found in InfoPortugal's map; total number of kilometers of roads that are not present on InfoPortugal's map (cartographic updating). Using vehicular sensing, we were able to build a road map comprising 421.82 km of roads. InfoPortugal's map has a total of 522.778 km of roads. Note that some of the roads of InfoPortugal's map, which are built based on aerial photography, correspond in fact to pedestrian-only zones and could never be detected based on vehicular sensing.

One of the most important results from the process of automatic road network extraction based on vehicular sensing consists in the identification of roads that are still non-existent in current maps. This aspect shows the ability of vehicular sensing to provide an inexpensive and highly accurate way of constantly performing cartographic updating. Our results present a correspondence of 82.94% between the kilometers of roads that were extracted and those from the base map. Hence, the remaining 17.06% represent the percentage of cartographic updating, i.e., the extraction of non-existing roads in the base map. Regarding the correspondence of our map with InfoPortugal's map, our results present a correspondence of 82.94% between the kilometers of roads that were extracted and those from the base map. Hence, the remaining 17.06% represent the percentage of cartographic updating, i.e., the extraction of non-existing roads in the base map.

The evaluation of the accuracy of the geometry of the extracted road network is done in a continuous manner: the average distance between the extracted roads and the corresponding roads in the base map is obtained through the computation of the area between the two, divided be the average length of the two geometric representations. This method presents a more accurate evaluation as compared to discrete based measurements. The resulting average distance was of 1,43 meters between the two representations of the same roads. The evaluation of the topology of the extracted road network is done through two main metrics: the number of nodes that have been extracted that match nodes in the base map; and the traffic direction of each extracted edge, compared to the direction of the associated edge in the base map. We were able to extract 596 nodes (road segment intersections) that matched with 458 nodes of the base map. Regarding the traffic direction on the matched road segments, vehicular sensing correctly identified 1158 of the 1206 road segments. The nodes which have no matching node on the base map do not necessarily represent false intersections. Such nodes can result from junctions of wide roads into more narrow segments, where the algorithm of automatic extraction usually created a non-existent node in the base map. Such nodes can also result from cartographic updating and the identification of intersections with non-existent roads in the base map. The last evaluation concerns the traffic rules between the generated roads. The extracted enforcements between connectivity of specific edges, stored in the direction table, were checked against the Do Not Enter signs of the base map, and reported an accuracy of 98%.

5   Vehicular Sensing Based on Windshield Cameras

With the introduction of cameras in production cars, and the wide dissemination of wireless inter-vehicle communication devices in all new vehicles in the foreseeing future, a potential application of vehicular sensing is the localization process of a particular vehicle, as the result of a broadcasted search warrant from some particular authority.

While these cameras have been designed to collect information to assist the driver and improve driving safety, with the support of a V2V communication infrastructure, all the in-car cameras disseminated in a city can provide a visual scope of tremendous proportions. A license-plate-based localization of vehicles, is an example of a delay-tolerant application that profits highly from the increased

Recognized Plates: 30-AE-08

Fig. 4. License plate recognition in a traffic jam situation

connectivity (radio and visual) of a set of mobile vision and communication enabled sensors. The automatic license-plate recognition is currently deployed in a number of fixed points, e.g. parking lots, gas stations and highway toll collectors, which can be globally connected to a police database of search warrants and instantaneously report the match. However, the visual scope or visual connectivity of such fixed sensors is clearly not comparable to the visual perception supported by in-car cameras distributed massively over the road network, which allow the localization of vehicles in large cities in a matter of seconds. Moreover, the vehicular sensor network empowers not only a distributed collection of data from the vehicles, but a more powerful distributed computation, that resorts to the processing power available in cars that are able to identify traffic signs or even pedestrians on the road. If care is taken to guarantee the secure and authorized usage of such an infrastructure, then vision-enabled vehicular sensor networks will possibily become critical monitoring tools.

In [14] we have described a car search protocol implemented on top of a vision-enabled vehicular newtork, where police cars issue search warrants based on license plates and the nodes of the network use the camera sensor to capture images, process them to recognize license plates, match the license plates against the search warrant and report the position through the Vehicular Ad Hoc Networks (VANET) until it reaches a police car 4. The objective was to understand how fast a query to this vision-enabled network, would return. This has many applications such as locating stolen vehicles, or lost children. In this work we focused on the first example.

We designed an experiment in which after the search warrant is issued, 4 minutes are allowed to find the stolen vehicle. Otherwise, it is considered that it was not found. In case the stolen vehicle leaves the map, or parks in a garage, it is considered that the vehicle escaped. Through Development of Intervehicular Reliable Telematics (DIVERT) traffic simulator [7], we have thoroughly evaluated this application using a realistic urban scenario (highly detailed map of the city of Porto) with thousands of vehicles, divided in several categories (police cars, vision-enabled cars, communicating cars, normal cars).We considered a very-sparse network of only 5000 vehicles, fixing the number of "police cars" at 50 (a realistic value in the city of Porto), and varied the percentage of "communicating cars" (10% and 20%), and of those, the percentage of "vision-enabled cars" (50% to 100%). We also considered diferent ranges of vision, that is the distance at which it is possible to "read" the license-plate (5 and 15 meters). The results are summarized in Table 1. For further details on the experiment, and results for different scenarios, please refer to [14].

The obtained results clearly are affected by the sparse network, which however may be accurate in representing the night period in a city like Porto, when more vehicles are stolen. Still, in a scenario where only one fifth of the vehicles are able to communicate, the results were very good and in at least 50% of the tests, the stolen vehicle was found within the allowed 4 minutes. These results can be extrapolated to other application such as the ones already mentioned, or others such as locating free parking slots in a city [5].

Table 1. Results of the locating stolen vehicles experiment

Rear and front plates results							
range of view	5m				15m		
# communicating sensors	500		1000		500		1000	
% vision enabled sensors	50%	100%	50%	100%	50%	100%	50%	100%
delay (seg)	143.53	143.28	141.16	136.41	144.58	125.59	119.80	111.15
not founded	57.5%	44.0%	40.1%	28.0%	46.0%	19.0%	18.0%	7.00%
escaped	41.0%	39.5%	40.0%	35.5%	38.5%	30.5%	29.0%	27.0%

6   Sensing Origin/Destination in a Taxi Network

Taxis are an important mean of transportation which offers a comfortable and direct transportation service. In the last decade, the real time GPS location systems became a key player on every taxi networking all over the world. All the vehicles are equipped with sensors continuously transmitting its GPS coordinates and instant speed, among others.

The streaming data provided for such network can highlight useful insights about the mobility patterns of an entire urban area over time and space like the following:

1. A cubic matrix origin/destination/time describing the mobility demand;
2. The seasonality of mobility problems such as non-forced traffic jams;
3. The characterization of the traffic flow patterns;

We focused into explore the insights described in the bullet 1 to improve the taxi driver mobility intelligence (i.e. to pick up more passengers and therefore, increase their profit). There exist just a few works on this specific topic [21,19,20,24]. Mainly, they suggest offline strategies (i.e. intelligent routing) to improve the passenger finding depending on time/space in scenarios where the demand is greater than the offer. However, there are cities with inverse scenarios (i.e. multiple companies with fleets larger than the actual demand competing between it selves) where just a fast and online strategy can actually speed the gains up.

In [26] we present a ubiquitous model to predict the number of services on a taxi network over space (taxi stand) for a short-time horizon of P-minutes. Based on historical GPS location and service data (passenger drop-off and pick-up), time series histograms are built for each stand containing the number of services with an aggregation of P-minutes. Our goal is to predict at the instant t how many services will be demanded during the period [t, t+P] at each existent taxi stand, reusing the real service count on [t, t+P] extracted from the data to do the same for the instant t+P and so on (i.e. the framework run continuously in a stream).

This model stands on three distinct pillars: 1) periodicity; 2) frequency and 3) a sliding time window. The first two correspond to two distinct forecast methods cleverly aggregated by the third one. The demand on taxi services exhibit, like other transportation means, a 1) periodicity (see Fig. 5) in time on a daily basis that reflects the patterns of the underlying human activity: so we used both seasonal and non-seasonal time varying Poisson models [17] to predict the service demand. When the frequency is distinct from the expected one (i.e. weekly), we have to handle it. The Autoregressive Integrated Moving Average (ARIMA) [3] models were used to such task by its well-known regressive properties. Finally, the models are ensemble using a 3) sliding time window: both models are used to build a weighted average where the weights are the models accuracy in the last X time periods (i.e. where X is a predefined time window that will slide forward constantly).

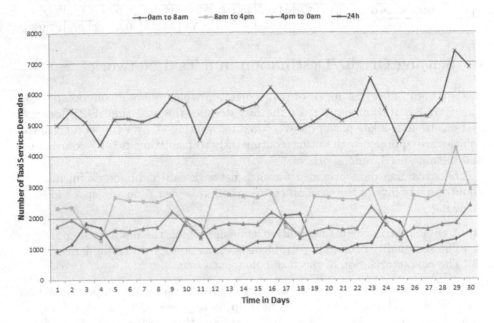

Fig. 5. One month taxi service analysis (total and per driver shift). There is a clear week periodicity in the data.

We applied this model to data from a large-sized sensor network of 441 taxi vehicles running on the city of Porto, Portugal. In this city, the taxi drivers must pick a route to one of the existing stands after a passenger drop-off. Our test-bed was a computational stream simulation running offline using 4 months data. The results obtained were promising: our model accurately predicted more than 76% of the services that actually emerged. This model will be used as a feature of a recommendation system (to be done) which will produce smart live recommendations to the taxi driver about which taxi stand he should head to after a drop-off.

7   Conclusions

The emergence of wireless networks in vehicles combined with the computation capabilities, mobility and energy-autonomy of modern in-vehicle sensor networks has the potential to create an ubiquitous platform for a wide range of sensing applications. Vehicular sensing allows to collect massive and varied information through sensors available at vehicles and to disseminate it using wireless communications, either using V2V/V2I or cellular networks. In this paper we have presented and demonstrated the feasibility/performance of three applications of vehicular sensing, namely automatic map construction and updating based on GPS traces, license-plate monitoring using windshield-installed cameras and taxi demand prediction per taxi stand over time, which allows a better understanding of urban dynamics.

References

1. Urbisnet: Urban pollution monitoring using a public transportation infrastructure for networked sensing (2012),
   http://users.isr.ist.utl.pt/~jpg/proj/urbisnet/urbisnet_main.html
2. Barcelo, J., Montero, L., Marques, L., Carmona, C.: Travel time forecasting and dynamic origin-destination estimation for freeways based on bluetooth traffic monitoring. Transportation Research Record: Journal of the Transportation Research Board 2175, 19–27 (2010)
3. Box, G., Pierce, D.: Distribution of residual autocorrelations in autoregressive-integrated moving average time series models. Journal of the American Statistical Association, 1509–1526 (1970)
4. Bruntrup, R., Edelkamp, S., Jabbar, S., Scholz, B.: Incremental map generation with GPS traces. In: Proceedings of IEEE Intelligent Transportation Systems, pp. 574–579. IEEE (2005)
5. Caliskan, M., Graupner, D., Mauve, M.: Decentralized discovery of free parking places. In: Proceedings of the 3rd International Workshop on Vehicular Ad Hoc Networks, pp. 30–39. ACM (2006)
6. Commission, E.: Commission Recomendation of 8 September 2011 on support for an EU-wide eCall service in electronic communication networks for the transmission of in-vehicle emergency calls based on 112 (eCalls). Official Journal of the European Union 54(L303), 46–47 (2011)
7. Conceição, H., Damas, L., Ferreira, M., Barros, J.: Large-scale simulation of v2v environments. In: Proceedings of the 2008 ACM Symposium on Applied Computing, pp. 28–33. ACM (2008)
8. Conceição, H., Ferreira, M., Barros, J.: On the Urban Connectivity of Vehicular Sensor Networks. In: Nikoletseas, S.E., Chlebus, B.S., Johnson, D.B., Krishnamachari, B. (eds.) DCOSS 2008. LNCS, vol. 5067, pp. 112–125. Springer, Heidelberg (2008)
9. Conceiçao, H., Ferreira, M., Barros, J.: A cautionary view of mobility and connectivity modeling in vehicular ad-hoc networks. In: IEEE 69th Vehicular Technology Conference, VTC Spring 2009, pp. 1–5. IEEE (2009)
10. Douglas, D., Peucker, T.: Algorithms for the reduction of the number of points required to represent a digitized line or its caricature. Cartographica: The International Journal for Geographic Information and Geovisualization 10(2), 112–122 (1973)
11. Edelkamp, S., Pereira, F., Sulewski, D., Costa, H.: Collaborative map generationsurvey and architecture proposal. In: Urbanism on Track Application of Tracking Technologies in Urbanism. Research in Urbanism Series, vol. 1. IOS Press, Amsterdam (2008)

12. ElBatt, T., Saraydar, C., Ames, M., Talty, T.: Potential for intra-vehicle wireless automotive sensor networks. In: 2006 IEEE Sarnoff Symposium, pp. 1–4. IEEE (2006)
13. Eriksson, J., Girod, L., Hull, B., Newton, R., Madden, S., Balakrishnan, H.: The pothole patrol: using a mobile sensor network for road surface monitoring. In: ACM MobiSys (2008)
14. Ferreira, M., Conceição, H., Fernandes, R., Reis, R.: Locating cars through a vision enabled vanet. In: 2009 IEEE Intelligent Vehicles Symposium, pp. 99–104. IEEE (2009)
15. Google: Self-driving car test: Steve mahan (2012), http://goo.gl/k5K9Q
16. Hu, S.C., Wang, Y.C., Huang, C.Y., Tseng, Y.C.: A vehicular wireless sensor network for CO_2 monitoring. In: IEEE Sensors, pp. 1498 –1501 (October 2009)
17. Ihler, A., Hutchins, J., Smyth, P.: Adaptive event detection with time-varying poisson processes. In: Proceedings of the 12th ACM SIGKDD International Conference on Knowledge Discovery and Data Mining, pp. 207–216. ACM, New York (2006)
18. Jiang, D., Delgrossi, L.: IEEE 802.11 p: Towards an international standard for wireless access in vehicular environments. In: IEEE Vehicular Technology Conference, VTC Spring 2008, pp. 2036–2040. IEEE (2008)
19. Lee, J., Park, G.-L., Kim, H., Yang, Y.-K., Kim, P., Kim, S.-W.: A Telematics Service System Based on the Linux Cluster. In: Shi, Y., van Albada, G.D., Dongarra, J., Sloot, P.M.A. (eds.) ICCS 2007. LNCS, vol. 4490, pp. 660–667. Springer, Heidelberg (2007)
20. Lee, J., Shin, I., Park, G.L.: Analysis of the passenger pick-up pattern for taxi location recommendation. In: International Conference on Networked Computing and Advanced Information Management, vol. 1, pp. 199–204 (September 2008)
21. Li, B., Zhang, D., Sun, L., Chen, C., Li, S., Qi, G., Yang, Q.: Hunting or waiting? discovering passenger-finding strategies from a large-scale real-world taxi dataset. In: 2011 IEEE International Conference on Pervasive Computing and Communications Workshops, pp. 63–68 (March 2011)
22. Li, X., Shu, W., Li, M., Huang, H.Y., Luo, P.E., Wu, M.Y.: Performance Evaluation of Vehicle-Based Mobile Sensor Networks for Traffic Monitoring. IEEE Transactions on Vehicular Technology 58(4), 1647–1653 (2009)
23. Lima, F., Ferreira, M.: Mining spatial data from GPS traces for automatic road network extraction (2009)
24. Liu, L., Andris, C., Biderman, A., Ratti, C.: Uncovering taxi drivers mobility intelligence through his trace. IEEE Pervasive Computing 160 (2009)
25. Mahmud, S., Alles, S.: In-vehicle network architecture for the next-generation vehicles (2005)
26. Moreira-Matias, L., Gama, J., Ferreira, M., Damas, L.: A predictive model for the passenger demand on a taxi network. In: Intelligent Transportation Systems Conference (to appear, September 2012)
27. Norton, H.N.: Handbook of Transducers. Prentice Hall (1989)
28. Rodrigues, J., Vieira, F., Vinhoza, T., Barros, J., Cunha, J.: A non-intrusive multisensor system for characterizing driver behavior. In: International IEEE Conference on Intelligent Transportation Systems, pp. 1620 –1624 (September 2010)
29. Schroedl, S., Wagstaff, K., Rogers, S., Langley, P., Wilson, C.: Mining GPS traces for map refinement. Data Mining and Knowledge Discovery 9(1), 59–87 (2004)

Framework for a Self-managed Wireless Sensor Cloud for Critical Event Management

Nithya G. Nair, Philip J. Morrow, and Gerard P. Parr

School of Computing and Information Engineering, University of Ulster, Coleraine, UK
gopalakrishnannair-n@email.ulster.ac.uk,
{pj.morrow,gp.parr}@ulster.ac.uk

Abstract. Wireless sensor networks (WSNs) can be widely used for managing various scenarios existing in social, industrial and numerous environmental management systems. They have been widely used in environmental monitoring and management applications and have also found application in disaster management scenarios. One of the greatest problems faced by the scientific community in organizing data collection through sensor networks in areas of disaster is the disorder and destruction brought about in the communication systems prevailing in such situations. In this paper, a scientific study of the various scenarios that could occur post-disaster and the various housekeeping functions each sensor node would adopt as part of the self management requirement is provided. We also present a sensor task management framework that could be implemented to provide a low energy consuming, reliable network for WSNs deployed for critical infrastructure management.

Keywords: self-managed wireless sensor cloud, critical event management.

1 Introduction

The demand for wireless sensor networks has increased in recent years due to advances in technology, which has led to extensive research into the field of critical infrastructure monitoring/management. One of the greatest problems faced by the scientific community in organizing data collection through sensor networks in areas of disaster is the disorder and destruction brought about in the communication systems prevailing in such situations like tsunamis, typhoons, earthquakes etc. With the increasing popularity of the emergent concepts of cloud technology, it would be quite beneficial to have the data recorded by wireless sensor networks deployed across a large area, made accessible from anywhere in the world.

Some of the most desired characteristics of such a network could be stated as follows: it performs energy efficient operations to increase the lifetime of the network and it possesses auto configuration capability as there may be a number of nodes that could join or leave the sensor cloud as the calamity sweeps across the region. When a WSN is deployed over a large geographical area, it assumes the format of a multi-hop communication network. It should also become data centric to avoid data loss due to node failures. The other characteristics that would be desirable are higher quality of

F. Martins, L. Lopes, and H. Paulino (Eds.): S-Cube 2012, LNICST 102, pp. 15–29, 2012.

service, higher fault tolerance, scalability of the network, lower power consumption/ better power management, better security, programmability, ease of maintenance and lower costs.

The focus of this paper is on Wireless Sensor Networks that are deployed to monitor and sense critical data. These could be event monitoring applications, for example: sensing seismic vibrations to warn of the possibility of a tsunami or landslide or flash floods and manage scenarios of post quake effects. Such WSNs could be deployed in an area with limited access and may be prone to drastic topological changes such as landslides and flood displacements. Since a sensor network is being utilized for critical applications, data being sensed would be of high importance. This requires the data being transmitted by the WSN to be highly reliable and error-free. The area to be sensed may be geographically inaccessible which would make it a top priority to make the network self-manageable and have an extended life time of operation.

A WSN deployed for monitoring purposes is susceptible to the situation of chaos as a result of the occurrence of a disaster. It would have to reconfigure using autonomic self-intelligent methods from a chaotic disarranged sensor distribution to form a robust network as was previously in place. This robust network is expected to carry out data collection and also provide reliable data transmission and delivery to a sink node. These functions have to be performed using the lowest energy consumption possible to extend the lifetime of the network. Hence, the important features a WSN monitoring a critical infrastructure would require are a) robust management protocols b) efficient power management and c) reliable data management and delivery.

The contribution of this project would be an efficient task management system with appropriate task suppression mechanisms that could contribute to the reliability of the deployed sensor network.

In this paper we are focusing on the specific issues dealt with when designing a framework for a self-managed wireless sensor network. The remainder of the paper is organized as follows: Section 2 provides a brief overview of related work done on the elements that contribute to the design of WSNs; Section 3 gives details on the various management aspects required in the design of a self-managed WSN; Section 4 gives an outline of the various scenarios that could arise in the case of a disaster and Section 5 concludes the paper and outlines further work that has to be carried out.

2   Related Work

Wireless sensor networks used for monitoring applications can be prone to various events like nodes dying, topology reconfiguration, loss of connection to the base station etc. Depending on various events that could occur during a WSN deployment, elements that could be said to contribute to the design of a WSN are protocols, data fusion / aggregation techniques, topology management and methods of deployment.

Protocols that are used in WSNs can be categorized as network / routing, transport and MAC protocols. There have been several network protocols designed to address the issue of connectivity establishment. According to the authors of [1], LEACH

(Low Energy Adaptive Clustering Hierarchy) is a popular choice among the various network protocols as it is flexible and self-adaptive by nature and uses a TDMA (Time Division Multiple Access) technique for transmission at the Management Node (Gateway) level. This makes it more energy efficient. Communication between nodes within a cluster is done through a CDMA (Code Division Multiple Access) to prevent interference between neighboring node transmissions. This protocol is said to be robust as node failures would not affect the overall network connectivity. The reason that this protocol would not be fit for implementation in the application mentioned here is because the initial setup phase of this protocol is quite energy consuming and has a large overhead when forming clusters. The deployment scenario discussed in this paper would require the network to be a multi hop communication infrastructure. LEACH is single hop communication protocol which would not be suitable for our purposes. It is understood that hierarchical/cluster based routing can be used in one-to-many and many-to-one communication scenarios. Hierarchical routing also helps reduce the overall energy consumption thereby increasing the lifetime of the network and providing scalability to the network [2]. These features make hierarchical routing protocols a favorable option to be used for the application mentioned here.

Due to limited bandwidth and the convergence nature of upstream data, two major issues to be dealt with would be packet loss and congestion within the WSN. Transport protocols are supposed to provide orderly transmission, flow and congestion control and QoS guarantee [3]. When dealing with WSN they need to provide a simple connection establishment process which has reduced transmission delays. Cross-layer optimisation would be an added advantage as the transport protocol would recognise if packet loss was caused due to congestion or route failure.

However, in [4] the authors indicate that TCP based protocols are not suitable for sensor networks as they do not consider energy conservation as a priority. The authors suggest that the Event-to-Sink Reliable Transport (ESRT) protocol supports congestion detection and also provides an acceptable level of reliability to the WSN and could be used for critical infrastructure monitoring.

When considering MAC protocols, random access protocols are considered suitable for WSN because of their self-organising nature. Since the project undertaken here is an event based application, duty cycle based and hybrid protocols could also be considered for use. ER-MAC [5] is an obvious choice as this protocol was designed for emergency response applications and unlike fixed protocols, provides scalability and flexibility to the network. The PACT algorithm [6] is another possible choice as it provides low latency and better QoS. It uses an adaptive duty cycle that depends on the traffic load thus increasing the network life time.

Data management in the case of WSNs would involve the task of acquiring data from sensors, storage of the collected data and efficient transmission of data to the end users [7]. It should be energy efficient, scalable and robust to failures as well. The sensors can be considered as distributed storage points which could be queried on demand. Efficient data transmission would require appropriate levels of data aggregation in order to preserve energy within the sensor node. Incorporating data fusion/aggregation techniques would provide enhanced system reliability, robustness due to redundant data being available, improved bandwidth utilization, lower energy consumption and extended network lifetime [8][9].

In [10] the authors explain topology management as the capability of the network to maintain connectivity across the network. Topology control is essential to increase the lifetime of a network by helping reduce the energy expenditure. It is also useful in reducing radio interference, improving the efficiency of MAC and routing protocols and also improves the robustness of the network and thereby the reliability of the network, quality of connection and also the coverage and have an influence on the scalability of the network. The authors in [11] suggest that these algorithms trade energy conservation for increased routing latency. They propose an algorithm for WSNs in failure-prone environments called adaptive Naps which conserves on energy and also is able work in a distributed manner thereby making it adaptive to node failures which was the disadvantage in the Naps algorithm.

Regarding deployment methods for a WSN, there are two types of deployments. They are deterministic and random [12]. A deterministic deployment is required when the cost of the sensor is high and their operation is significantly influenced by the location, e.g.: underwater applications (imaging sensors) or indoor applications. Random deployment is used where the cost of nodes is not of high priority and also where the environment is harsh, e.g.: battle field or disaster areas. The factors on which node deployment depends are: area coverage, network connectivity and longevity and data reliability. Other deployment strategies available depend on the functions of the node as well. In order to achieve maximum area coverage, random deployment may be a feasible option which is widely adopted in applications like target tracking etc.

Thus, we can say that the protocols that could be adopted for use within this project can be summarised using the protocol stack described by the authors in [3] as in Figure 1:

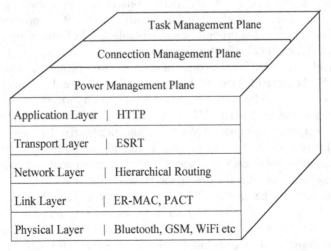

Fig. 1. Protocol stack adapted from [3]

There are several case studies where the authors provide WSN architectures for specific applications. In [13] the authors discuss an architecture that is self configurable in terms of topology reconfiguration in an energy efficient manner. This particular architecture does not talk about data management which is also vital in applications like monitoring of critical infrastructures. The authors of [14] describe an architecture that deals with a low power network management but have not given enough importance to data reliability and management. Another case study for a WSN used for disaster management is given in [15]. This network is used to manage rescue operations after occurrence of a large scale disaster. Here the nodes are randomly deployed by individuals at their home, office etc. In this network the end user is mobile and so can collect data from the nodes without needing to have an extensive setup for connection management and data management etc.

3   Task Monitoring Framework for WSNs

A deployed WSN may experience events like topology changes or nodes dying or malfunctioning etc. The scenarios that arise as a result of these events are explained in Section 4. An 'event' could be a change in any parameter used by the node for initial setup. For example, movement of the node, loss of connection and depletion of energy or memory reserve. The 'event' could also be described as an external event that would be picked up by a sensor on the node. It could be any event like change in temperature, vibration, detection of movement etc. The events mentioned above would require the WSN to reorganize and adopt functionalities that would attempt to provide network connectivity to a majority or all nodes present in the WSN. In order to achieve this, various management functions have to be executed. The management of a WSN could be influenced by sensor tasks that are needed to be carried out within the network.

These tasks could be classified according to the modes of operation of the sensor, the interaction among sensor nodes, mobility of the sensor which in turn would deal with the resource management within the node, and also the priority of the tasks being executed.

3.1   Sensor Tasks

Figure 2 provides a classification of the various tasks that could be performed by the node. Modes of operation of a sensor node could be either autonomous (stand alone) or connected (networked to base station or adhoc). Events may occur that could cause the connectivity within the network to change and hence the node may find itself in various situations such as:

> ➤ The sensor has a connection to the base station (networked)
> ➤ The sensor is part of a group/network of sensors that would have lost contact with the base station (ad-hoc mode)
> ➤ The sensor is alone, disconnected from the network (i.e. in standalone mode)

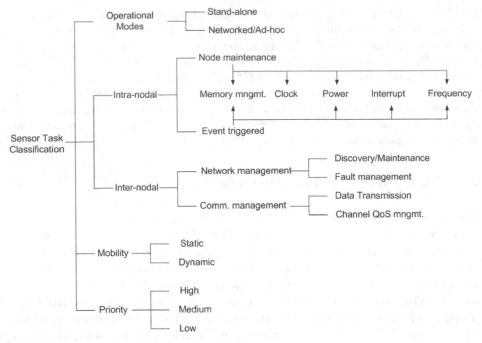

Fig. 2. Sensor task Classification

Depending on the interaction of nodes when the execution of a task takes places, they could be divided into intra-nodal and inter-nodal tasks. Intra-nodal tasks could further be divided into node self management / maintenance tasks and event triggered tasks. The self-management/maintenance tasks carried out by the node are memory management, clock synchronization tasks, power management and task-frequency management. The event triggered tasks would overlap with most of the self-management tasks performed by the node but would also include interrupt management tasks.

Inter-nodal tasks require the node to have interaction with other nodes in the form of polling and information transfer etc. These are further divided into network management tasks and communication management tasks. The network management tasks involve tasks required to be performed for discovery and formation of the sensor network and also maintenance of the network. The network management tasks also deal with the fault management of the sensor network setup. The communication management tasks deal with the data transmission and also the channel QoS management.

The sensor tasks could further be classified based on the mobility of the sensor i.e. the resource management tasks that would be required if a node is static or in a dynamic environment. The resource management tasks would overlap with the intra-nodal tasks as it would be related to tasks that maintain the node thereby attempting to extend the node's lifetime.

The sensor tasks could also be classified according to priority i.e. high, medium and low. This is provided to put in place a task suppression system which would be essential to avoid overloading the node CPU and also wasting the limited resources available to the node. It would also provide a level of legitimacy to the alerts that would be created by the sensor network.

3.2   Task Management Groups

The various sensor tasks have been discussed above and can be further categorized into management groups. Within our framework we propose the following systems / groups: software management, power management, sensor assessment, buffer management, data management and task frequency management groups. A $P_{i,j}$ matrix is used to represent the various parameters within each management system where i is the management group and j is the state.

Software management refers to the various states that a sensor node could adopt depending on the situation it finds itself in. Three states can be defined for node operation as follows:

Active state $(P_{1,1})$:	All systems are functioning at full power
Idle/Standby $(P_{1,3})$:	The node is in low power state where the transceiver is turned off to save power.
Sleep/ Off $(P_{1,2})$:	The node is completely inactive for a set time period. It would poll for anomalies in between the idle cycles to decide if it needs to wake up and change state or not.

The Active state is executed for data processing and transmission. The Sleep/Off state is executed when the node is not in use for a particular cycle. The Idle/ Standby state is adopted when the node is sensing but no data processing or transmission is needed. The energy levels that are consumed during these states vary and the switching between these states could be used in the future to formulate an optimal energy conservation technique.

Power management is required by the sensor node to conserve battery power by means of regulating node functions thereby extending the lifetime of the sensor network as a whole. Sensor node battery level thresholds could be defined before deployment. The thresholds could be defined as T1 for normal level and T2 for critical level (Fig. 3).

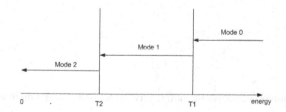

Fig. 3. Power thresholds and modes for power management

Depending on the thresholds we could define three modes of operations. They are:

> **Mode0** $(P_{2, 1})$: sensor node functioning.
>
> **Mode1** $(P_{2, 2})$: send a signal to gateway about "time to extinct" period to cease message bridging functions undertaken.
>
> **Mode2** $(P_{2, 3})$: sensor node in critical level and so emergency measures adopted.

'Time to extinct' period is the time that would be taken for the node's battery level to reach critical level after which most of the node's regular functions would be put on hold.

Sensor Assessment is a house keeping function used by the node to decide whether the data obtained from the sensors is reliable i.e. if the error tolerance is within the acceptable band. The band is to be pre-determined before deployment. Three levels that can be used to determine if the data received is reliable or not, are:

Sensor intact $(P_{3, 1})$: data collected during each cycle is within the acceptable margins.

Sensor lost $(P_{3, 2})$: data received is garbled continuously or no data received by the node from the sensor.

Sensor doubtful $(P_{3, 3})$: if there are occurrences of garbled data within a set of observations the sensor state would be at a level where the sensing function would be considered doubtful

Buffer Management deals with memory management i.e. the management of memory levels within a sensor node. Data management strategies on the received and stored data could be adapted when a particular level is reached in the memory storage. This level could be calculated according to a *relinquishing ratio* (Fig. 4) which depends on the rate at which data is sent to a safe location from the current node. This could also be set at the pre-deployment period, depending on the characteristic of the predicted use and event occurrences.

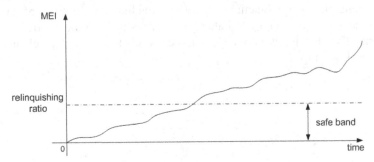

Fig. 4. Depiction of relinquishing ratio relative to Memory engagement index (MEI)

Memory engagement index (MEI) is defined as the ratio of the buffer occupied by data and the total size of buffer.

The *relinquishing ratio* is the MEI value which distinguishes levels below it as the safe band. This ratio could be adjusted according to the different states in which the sensor would operate.

Data management is closely related to buffer management in terms of *data sieving*. Data sieving refers to data quality management of the sensed data. Data quality could be divided into three bands. i.e., Positive faults, likely to be a fault and acceptable data.

$$\text{Data quality bands} \begin{cases} \text{DQ1} => \text{positive fault } (P_{4,1}) \\ \text{DQ2} => \text{likely to be a fault } (P_{4,2}) \\ \text{DQ3} => \text{acceptable data } (P_{4,3}) \end{cases}$$

When the data received by the sensors falls within DQ1 or DQ2 it is discarded and if it is within DQ3 it would be saved within the node memory, if MEI is favorable. This affects the buffer management and energy consumption. It contributes to energy conservation by avoiding unnecessary processing of data that does not fall in the acceptable range.

Data management also includes housekeeping functions for *data communication* from the sensor node to the base station. A healthy communication would normally work on a fixed range of transmission frequencies. When an event occurs and loss of communication takes place, the node can switch over to recovery mode i.e. to achieve maximum energy conservation and data preservation. In the case of a post disaster communication channel loss, we could adopt an exponentially behaving *rescue-cling algorithm* (Fig. 5). The purpose of this algorithm would be to enable a ping sequence for other nodes with gradually decaying polling frequency on repeating failures to avoid excess use of power. In case an acknowledgement is received, the node would maintain the normal repetition rate at which it sends messages from then on. If no acknowledgement is received then the frequency with which polling is done decreases exponentially.

Fig. 5. Frequency change when using *rescue-cling* algorithm

Data aggregation is also an operation which forms part of data management which occurs during failure mode. When the sensor is isolated, a failure mode flag is set and a dynamic interval sampling frequency (i.e. sensing frequency) may be adopted. In the case of a gateway, compression techniques may be applied to the data that is being transmitted to the base station.

Task Frequency Management: All repetitive processes within the sensor nodes will be governed by certain frequency properties. Major domains for this house keeping function are the sensor data sampling functions and the data transmission cycles. Variations in these rates of operations depend on the various situations in which the sensor nodes find themselves.

3.3   Sensor Task / Management Relationships

As discussed above, various management groups have been defined along with several parameters. Most of the sensor tasks discussed earlier in the section overlap between the various management groups. Table 1 provides the various sensor tasks and management groups and the overlap of tasks with the management groups.

Table 1. The placement of the various sensor tasks in the different groups

Sensor Tasks / Management Groups		Software	Power	Buffer	Data	Sensor Assessment	Frequency
Modes of Operation	Stand alone	✓	✓	✓	✓		✓
	Networked/ Ad-hoc	✓	✓			✓	
Intra-node	Memory			✓	✓		
	Clock						✓
	Power		✓				
	Interrupt	✓		✓		✓	✓
	Frequency		✓			✓	✓
Inter-node	Topology Discovery/ Maintenance	✓	✓		✓		✓
	Fault Management				✓		✓
	Data transmission		✓	✓	✓	✓	✓
	Channel QoS management		✓	✓	✓	✓	✓
Mobility	Static	✓	✓	✓	✓		✓
	Dynamic	✓	✓	✓	✓	✓	✓
Priority		✓	✓	✓	✓	✓	✓

A 'tick' indicates that a sensor task as listed on the left side of the table has a relationship with a management group listed on the top of the table. Consider the case of an interrupt task. Within a node, several interrupts maybe present that could influence the software of a node, buffer management and frequency management. The sensor assessment would have an influence on the interrupt tasks i.e. by determining if the sensor is functioning or not, an interrupt could be executed which would in turn influence one of the other management groups it is related to.

4   WSN Scenarios Following an Event Occurrence

As discussed in Section 3, there are various tasks that could be executed when an event occurs. This is in relation to the various situations in which the node would have to operate i.e. the node could find itself disconnected from the network or it could find itself connected to a cluster of nodes that has lost connection to the network.

We investigate a number of different scenarios that may be dealt with by a wireless sensor network when it adopts operations similar to adhoc wireless networks [16]. When an event occurs, the network infrastructure that was previously built could get partially or fully disorganized. This could result in a loss in communication/coverage of an area leading to loss of vital data that is required for analysis by the user.

An ideal world situation is considered initially to provide a baseline for comparison with the subsequent scenarios.

4.1   Ideal World (Scenario 1)

In an ideal scenario, as illustrated in Fig 6, there are no factors that could lead to loss in performance or reduce the reliability of the deployed wireless sensor network.

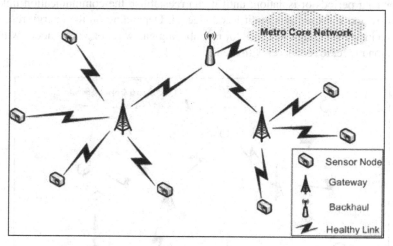

Fig. 6. Illustration of an Ideal World Scenario

There would be no possible loss of connectivity, no shortage of memory for vital data storage within the nodes, and energy within the nodes would not be exhausted. Thus, node failure due to energy depletion would not occur. The sensors that form part of the WSN could be deployed randomly or in a pre-deterministic manner and are connected to the sink or base station for operations. In the ideal world or near ideal world scenario, disruption to communication links would not occur and also minimal power management and data management techniques need to be adopted. Polling frequency can be varied as and when required.

When an event is triggered, the polling frequency is increased initially to determine if the event was a random occurrence or not. If an event trigger is received consistently over a period of time then an alarm is sent upstream to the core network and data is gathered for further analysis by the users.

4.2   Real World Scenarios

When dealing with real world scenarios, anything can go wrong. We look at two main types of scenarios which can be divided further into two subsections according to the housekeeping function they have to adopt to form a robust reconfigured network. Each of these scenarios give rise to a number of issues that must be considered for the continuous operation of the wireless sensor network and also extending the lifetime of the nodes and in turn the network as a whole. The issues that arise in the real world scenarios, mainly topology reconfiguration, power management and data management are also discussed.

Scenario 2:   Dynamic Topology Where There Is a Break in the Connection between the Node and Gateway but the Connection to Backhaul Still Exists (Fig 7)

Case (i): A sensor node loses its connectivity to the gateway/base station due to change in network structure. There are several housekeeping functions that would need to be undertaken. It would have to reduce its polling frequency to conserve power and storage space for long periods of isolation, until it can reestablish the communication link with the gateway and transmit the data it has collected. Depending on the energy reserves it could scan its surroundings, to check for neighboring gateways or sensor nodes within its transmission range, to reconnect with the network.

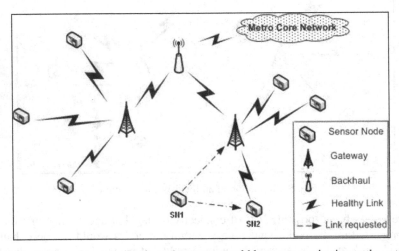

Fig. 7. Illustration of a scenario where the sensor would lose communication to the network

When an event occurs, the isolated sensors would increase their polling frequency and store the sensed data in the expectation that the link to the network would be restored at a later time. The situation that the node could experience in the isolated state would be depletion of memory. If such is the case, the oldest data could be deleted i.e. overwritten to store the newer data. Another option is to prioritize the data stored and the lower priority data could be overwritten if a high priority data is sensed and recorded. The priority levels of the data would have to become user defined using certain means of threshold setting etc.

Data aggregation would be necessary for the transmission of the data collected as it could be sometime before the communication link is re-established with the network and there may be lots of data to be transmitted. Attempting to transmit all the data without performing aggregation would result in depletion of the nodes energy reserves. It would consume the energy and storage reserves of the gateway or node with which it would reestablish connection as they would have their own data to transmit and also other sensors connected to them as well.

When considering the energy aspect of the node, when dealing with an emergency situation, if the energy reserve in the node is severely diminished, it would forgo the connection to the gateway and opt for a shorter range communication (Bluetooth etc. or IR) as and when it finds a suitable neighbor node.

Case (ii): The housekeeping functions that have to be adopted by a sensor node and the corresponding gateway when an isolated node gets into transmission range and requests a link to send the data gathered upstream towards the sink.

In the case when an isolated sensor node (SN1) would request a connection to the network through either a gateway or a sensor node connected to the network (SN2), there are certain housekeeping functions that need to be performed by the gateway or SN2 to accept SN1 into the network. Power management would be of importance if the connection is being established between SN1 and SN2 as SN2 will have to maintain connection with the gateway as well as cater to SN1 henceforth. Data aggregation would be required as SN2 would have its own gathered data to send upstream to the sink. If SN1 is able to establish connection with the gateway then the gateway would have to do further aggregation of the received data to accommodate the data received from SN1 and also let the other nodes in its vicinity know about the new addition i.e. SN1 to the network.

Scenario 3: The WSN Is Subjected to a Dynamic Topology Where the Connection to the Sink Is Lost to a Gateway (Fig 8)

Case (i): When a gateway loses connectivity with the sink/ backhaul or the bridging gateway. This case is slightly similar to the one where a sensor node loses its connection to the gateway. The main difference here is the level of aggregation required when transmitting the sensed data upstream towards the sink.

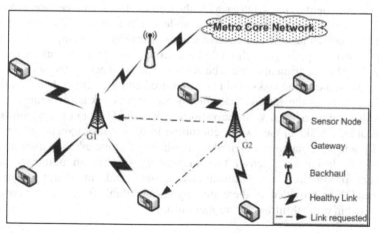

Fig. 8. Illustration of the scenario of a gateway attempting to reestablish connection to the backhaul through a bridging gateway or a sensor node

In this scenario when an event occurs and a gateway loses connectivity to the network, data management and in turn energy management becomes an issue. Data aggregation would have to be performed at a higher rate at the sensor node level and gateway level. Based on the residual energy of sensor nodes, some of them would be assigned the task of scanning their vicinity for other sensor nodes or gateways while others would go on with their task of sensing.

Case (ii) The housekeeping functions that have to be adopted by a sensor / gateway (G1) when a gateway with no communication link (G2) to the sink requests to reestablish a connection to the sink.

In this case, G2 request a link to either G1 or a child sensor node of G1. As with scenarios discussed earlier, G1 or a sensor node within the cluster of G1 would have to perform power management to compensate for the addition to the network. Data aggregation would have to be at a higher level as data of an entire subnet would have to be transmitted upstream through G1.

From the scenarios discussed here, the main issues identified are power management in terms of connectivity management and data management. In a post-event situation, when certain links to backhaul network is broken, the scattered network could reconfigure and form adhoc clusters which at some stage could reestablish connection to backhaul network. When connection is reestablished with sink, data management would have to be taken into consideration at that stage. Therefore, to build a WSN framework that is robust and energy efficient, protocols would have to be adopted that would need to provide connectivity management and data management with energy efficiency as an important feature.

5   Conclusion and Further Work

We are proposing a framework for a task monitoring system that provides details on the various sensor tasks that are executed when an event occurs. We discuss the different management groups into which the sensor tasks can be organized. Also discussed are the various scenarios that a node could find it-self in after an event occurs. An experimental test bed with Libelium Wapmotes is being used to create a multi-hop network topology which could be used to emulate the scenarios mentioned in Section 4. The experiments would be set up such that each of the scenarios could be created and some of the tasks could be monitored and thus create benchmarks for a metric of the costs for the various tasks for any sensor network applications.

The costs that are associated with the various tasks could be: 1) energy consumption when executing a task, 2) bandwidth consumption during task execution, 3) load on the CPU for a task, 4) memory required by a task, and 5) time taken to execute a task

Along with designing a sensor task monitoring system, an optimised network management protocol, a data management system and an energy management protocol need to be provided as there are very few available for a sensor network that would be used in critical infrastructure monitoring.

Acknowledgement. We would like to acknowledge the University of Ulster for providing a VCRS studentship as part of this IU-ATC (India-UK Advanced Technology Center of Excellence in Next Generation Networks) project.

References

[1] Frye, L.: Network management of a wireless sensor network, pp. 1–13 (2007)

[2] Lotf, J.J., Hosseinzadeh, M., Alguliev, R.M.: Hierarchical routing in wireless sensor networks: a survey. In: 2010 2nd International Conference on Computer Engineering and Technology, pp. V3-650–V3-654 (2010)

[3] Zheng, J., Jamalipour, A.: Wireless sensor networks: a networking perspective, vol. 2008, p. 489. Wiley-IEEE Press (2009)

[4] Wang, C., Sohraby, K., Hu, Y., Li, B., Tang, W.: Issues of transport control protocols for wireless sensor networks. In: Proceedings of 2005 International Conference on Communications, Circuits and Systems, pp. 422–426 (2005)

[5] Sitanayah, L., Sreenan, C.J., Brown, K.N.: ER-MAC: A Hybrid MAC Protocol for Emergency Response Wireless Sensor Networks. In: 2010 Fourth International Conference on Sensor Technologies and Applications (SENSORCOMM), pp. 244–249 (2010)

[6] Ali, M., Böhm, A., Jonsson, M.: Wireless Sensor Networks for Surveillance Applications – A Comparative Survey of MAC Protocols. In: 2008 The Fourth International Conference on Wireless and Mobile Communications, pp. 399–403 (2008)

[7] Cantoni, V., Lombardi, L., Lombardi, P.: Challenges for Data Mining in Distributed Sensor Networks. In: 18th International Conference on Pattern Recognition (ICPR 2006), pp. 1000–1007 (2006)

[8] Aguilar-Ponce, R., McNeely, J., Baker, A., Kumar, A., Bayoumi, M.: Multisensor Data Fusion Schemes for Wireless Sensor Networks. In: 2006 International Workshop on Computer Architecture for Machine Perception and Sensing, pp. 136–141 (September 2007)

[9] Chen, Y., Shu, J., Zhang, S., Liu, L., Sun, L.: Data Fusion in Wireless Sensor Networks. In: 2009 Second International Symposium on Electronic Commerce and Security, pp. 504–509 (2009)

[10] Gengzhong, Z., Qiumei, L.: A Survey on Topology Control in Wireless Sensor Networks. In: 2010 Second International Conference on Future Networks, pp. 376–380 (2010)

[11] Frye, L., Bigrigg, M.W.: Topology Maintenance of Wireless Sensor Networks in Node Failure-prone Environments. In: 2006 IEEE International Conference on Networking, Sensing and Control, vol. 15213, pp. 886–891 (2006)

[12] Younis, M., Akkaya, K.: Strategies and techniques for node placement in wireless sensor networks: A survey. Ad Hoc Networks 6(4), 621–655 (2008)

[13] Asim, M., Yu, M., Mokhtar, H., Merabti, M.: A Self-Configurable Architecture for Wireless Sensor Networks. In: 2010 Developments in E-systems Engineering, pp. 76–81 (September 2010)

[14] Pogkas, N., Karastergios, G.E., Antonopoulos, C.P., Koubias, S., Papadopoulos, G.: Architecture design and implementation of an ad-hoc network for disaster relief operations. IEEE Transactions on Industrial Informatics 3(1), 63–72 (2007)

[15] Cayirci, E., Coplu, T.: SENDROM: Sensor networks for disaster relief operations management. Wireless Networks 13(3), 409–423 (2006)

[16] Gopalakrishnan Nair, N., Morrow, P.J., Parr, G.P.: Design Considerations for a Self-Managed Wireless Sensor Cloud for Emergency Response Scenario. In: The 12th Annual Post Graduate Symposium on the Convergence of Telecommunications, Networking and Broadcasting, PGNet 2011 (2011)

Middleware Mechanisms for Agent Mobility
in Wireless Sensor and Actuator Networks

Nikos Tziritas[1,2,3], Giorgis Georgakoudis[1,2], Spyros Lalis[1,2], Tomasz Paczesny[4],
Jarosław Domaszewicz[4], Petros Lampsas[1,5], and Thanasis Loukopoulos[1,5]

[1] Center for Research and Technology Thessaly,
Technology Park of Thessaly 1st Industrial Area, 385 00 Volos, Greece
[2] Department of Computer and Communication Engineering,
University of Thessaly,
Glavani 37, 38221 Volos, Greece
{nitzirit,ggeorgak,lalis}@inf.uth.gr
[3] Shenzhen Institutes of Advanced Technology,
Chinese Academy of Sciences, Shenzhen, 518067, China
nikolaos@siat.ac.cn
[4] Institute of Telecommunications,
Warsaw University of Technology,
Nowowiejska 15/19, 00-665, Warsaw, Poland
{t.paczesny,domaszew}@tele.pw.edu.pl
[5] Department of Informatics and Computer Technology,
Technological Educational Institute of Lamia,
3rd km. Old Ntl. Road Athens, 35100 Lamia, Greece
{plam,luke}@teilam.gr

Abstract. This paper describes middleware-level support for agent mobility,
targeted at hierarchically structured wireless sensor and actuator network applications. Agent mobility enables a dynamic deployment and adaptation of the
application on top of the wireless network at runtime, while allowing the middleware to optimize the placement of agents, e.g., to reduce wireless network
traffic, transparently to the application programmer. The paper presents the design of the mechanisms and protocols employed to instantiate agents on nodes
and to move agents between nodes. It also gives an evaluation of a middleware
prototype running on Imote2 nodes that communicate over ZigBee. The results
show that our implementation is reasonably efficient and fast enough to support
the envisioned functionality on top of a commodity multi-hop wireless technology. Our work is to a large extent platform-neutral, thus it can inform the design
of other systems that adopt a hierarchical structuring of mobile components.

Keywords: wireless sensor networks, middleware, mobile agents, embedded
systems, performance evaluation, Imote2, ZigBee.

1 Introduction

In the POBICOS project [11] we have produced a platform aimed to simplify the development and deployment of monitoring and control applications for the home and office

F. Martins, L. Lopes, and H. Paulino (Eds.): S-Cube 2012, LNICST 102, pp. 30–44, 2012.
© Institute for Computer Sciences, Social Informatics and Telecommunications Engineering 2012

environment, which exploit regular objects with embedded sensing, actuating and wireless communication capabilities. Objects do not have any application-specific code pre-installed and are agnostic about the applications that might run on them. Each application is injected into the network (referred to as object community) using a special device (the application pill), which stores the code of the application and serves as its controller [6]. To start the application, the user simply pushes a button on the pill, letting the middleware deploy and execute the application on the object community.

POBICOS applications are programmed as a set of cooperating components, called agents. Agents are mobile in the sense that they can be instantiated on remote objects and can be migrated between objects, at runtime. Agent mobility is central to achieving non-trivial functionality. Firstly, it enables a flexible deployment of the application code in the object community, by placing individual agents directly on the objects that provide the required (computing, sensing, actuating) resources. Secondly, it allows the programmer to dynamically control the type of agents that execute in the object community, depending on the application's internal state. This in turn can reduce the amount of code that needs to be kept in the main memory of embedded nodes at any point in time, especially in the presence of several concurrently running applications. Thirdly, the middleware can migrate agents between objects in order to perform certain optimizations, e.g., to reduce the traffic over the wireless network. Unlike in many other embedded agent systems, agent mobility is transparent for the programmer who does not have to discover (suitable) objects or to deal with the placement of agents on objects in an explicit fashion.

This paper describes the protocols and mechanisms that were developed to support agent creation and agent migration in POBICOS. It also provides an experimental evaluation based on a prototype implementation of the middleware on Imote2 nodes that communicate over ZigBee. The results provide valuable insight into the overhead and performance of the agent mobility operations on top of a popular multi-hop wireless technology, showing that they are reasonably efficient and fast enough to support the envisioned functionality. Notably, this work is to a large extent orthogonal to the POBICOS platform: as explained in the next section, the basic underlying assumption is that agents are arranged as a tree according to their parent-child relationship. Hence the presented middleware support and performance trends can inform the design of other systems which adopt a hierarchical structuring of mobile components.

The rest of the paper is structured as follows. Sec. 2 provides an overview of the application model. Sec. 3 describes the implementation of the agent mobility protocols and mechanisms in our middleware. Sec. 4 analyzes their performance, while Sec. 5 puts the costs and benefits of agent mobility in context of an indicative application scenario. Sec. 6 discusses related work. Finally, Sec. 7 concludes the paper.

2   Application Model

The POBICOS application model evolved from that of the ROVERS system [3]. An application is designed as a collection of cooperating agents, with each agent being dedicated to a specific, perhaps very simple, task. In the spirit of hierarchical control systems [9], agents are organized in a tree. Leaf agents interact with the physical environment by acquiring information or effecting change through the sensors and actuators embedded into the objects of the community. The rest of the agents in the application tree implement higher-level aggregation, processing and control tasks, using only general-purpose computing resources (CPU and memory). Agents communicate via message passing. Being consistent with the hierarchical approach, an agent can exchange messages only with its parent and children.

Fig. 1a shows the structure of a simple application that turns off lights when there is no user activity (of course, applications can be more complex). The root agent (R) employs a user presence inference agent (I), which relies on multiple user activity agents (A) to detect user activity (and to infer inactivity). The root also employs multiple light agents (L) to switch off lights when the user presence inference agent (I) reports user absence, and a notification agent (N) to inform the user about this action.

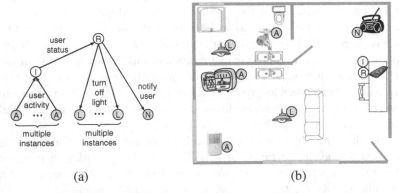

(a)   (b)

Fig. 1. (a) Tree structure of a simple light control application; (b) A concrete deployment of the light control application on top of an indicative object community

The agent tree is formed, at runtime, in a top-down fashion. The root is automatically created by the middleware on the application pill when the application is started. All other agents are created under the control of the application according to the desired tree structure. The placement of agents in the community is performed by the middleware, with no direct involvement of the programmer, based on the objects that are available. In the deployment shown in Fig. 1b, user activity agents are created on the motion detector, refrigerator and water tap, because these objects can serve as user activity sensors. Light agents are created on all light sources, while the user notification agent is created on the radio, which can issue messages to the user. The middleware can move non-leaf agents between objects in a transparent way. For instance, the user presence inference agent could be migrated from the application pill on the motion detector, to communicate locally with the respective user activity agent. Migration is supported only for non-leaf agents because they are object- and location-neutral by design; leaf agents remain on the nodes where they are created.

It is worth noting that parts of the agent tree can be instantiated and destroyed spontaneously, long after the application has been deployed. As an example, the light control application could create light agents only when the decision is taken to turn the lights off, and then remove them once they perform their task (as opposed to creating them "statically" at startup). Of course, it is up to the programmer to decide if such dynamic changes in the application tree are meaningful.

3   Implementation of Agent Mobility

The middleware components involved in the implementation of agent mobility are shown in Fig. 2. The core functionality is provided by the Agent Manager, which invokes the Agent Runtime to check resource availability, as well as to initialize, run, suspend and resume agents. Agent binaries are downloaded via the Code Transport, employing a stop-and-wait protocol and a cache to avoid fetching the same code repeatedly over the network. The Network Abstraction offers a generic datagram interface, used by both the Agent Manager and Code Transport.

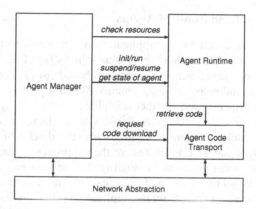

Fig. 2. Key middleware components and interactions for supporting agent mobility

Our prototype is developed for TinyOS v2.1 running on Imote2 nodes [8] at 104MHz. Wireless communication is done via an external ZigBee modem from the Z430-RF2480 demo kit of Texas Instruments [14]. The Network Abstraction component breaks datagrams into ZigBee packets and implements its own software-based acknowledgement and retransmission scheme (relying on ZigBee for packet routing). The middleware is portable, assuming support for TinyOS; obviously, the Network Abstraction must be adapted to the underlying networking technology. Thanks to a system component that provides transparent access to external memories (e.g., Flash), the minimal RAM requirements are below 8KB, making it possible to target more resource-constrained devices (even though access to certain middleware data structures would be slower, we believe that the middleware would still work acceptably).

3.1   Micro-agent Code, Execution and State

Nodes provide a platform-neutral runtime on top of which agents execute. The VM is based on the 8-bit AVR architecture [1]. Agents are written in C and the respective binaries are generated via the standard AVR-GCC tool-chain. The binaries are then processed using a special tool to bind into the POBICOS-specific primitives and reduce their size [12].

Agent execution is purely event-driven (agents do not have threads of their own). The Runtime puts events issued to agents in a FIFO queue, and executes the respective handlers in a non-preemptive fashion. Agents are migrated only between handler executions, when the stack is empty and the VM CPU is not being used, which greatly simplifies the respective suspend-resume process. Also, agents do not use dynamic memory, so the runtime state that needs to be transferred over the network when a migration takes place is just the agent's static data.

The Agent Manager maintains additional data for each local agent, namely entries for its children and any pending creation requests. In case of a migration, this information must also be sent to the destination along with the agent's state.

3.2   Creation of Leaf and Non-leaf Agents

Agent creation requests issued by the application are processed in an asynchronous fashion. The process for creating a leaf agent is as follows (see Fig. 3a).

To find nodes that can serve as hosts, a probe is broadcast (1) carrying information about the agent's type and resource requirements. When a probe arrives to a node, a resource check is performed to see whether it is able host such an agent (2). A reply is generated (3) only if this check succeeds. Replies are collected (4) and sorted based on how well nodes match the agent's requirements (the details of this matching are beyond the scope of this paper). Candidates are then approached one at a time.

A node is asked to create an agent by sending it a creation request (5). On receipt of such a request, to avoid races, the node repeats the same checks as for a probe (6). Then, it downloads the agent code from the application pill (7), creates a new instance (8), recording the sender of the request as the agent's parent, and sends back a reply with the agent's identifier (9). If the reply is positive, a new child entry is added to the parent (10). Else, if the reply is negative or no reply arrives within a timeout period, the next candidate is considered.

The creation of non-leaf agents works in a similar way. However, host discovery is performed only if the local host cannot host the agent, and candidates are contacted in the arrival order of their replies. The rationale is that since non-leaf agents are object-neutral it is reasonable to place them, at least initially, close to their parent.

3.3   Migration of Non-leaf Agents

The algorithm for deciding about the migration of a non-leaf agent in order to reduce the wireless traffic is described in [15]. The idea is to locally record the agent's messaging activity with its parent and children, and to move the agent towards the center of gravity, i.e., over the link that accounts for more than half of the total load. We have implemented the k-hop variant of the algorithm, which assumes knowledge about the routing structure within a k-hop radius and can pick migration destinations in this range. In ZigBee tree networks, where node addresses reflect the routing topology [10], this information can be gained without any extra communication.

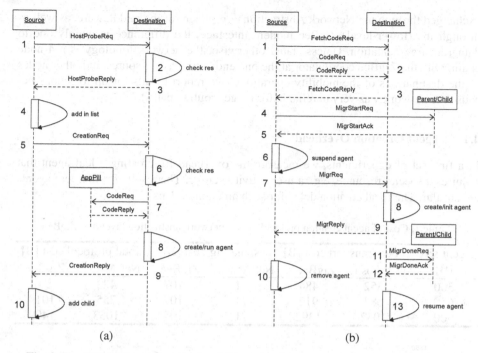

Fig. 3. Simplified (a) agent creation and (b) agent migration message sequence diagrams

Once a migration decision is taken, the process is as follows (see Fig. 3b). First, the destination node is asked to download the code (1-3). The code is fetched from the node that initiates the migration (which, obviously, has the binary). Then, the hosts of the agent's parent and children are notified (4) to buffer messages addressed to it, but also to prohibit concurrent migrations (parents have precedence over their children). When all acknowledgements arrive (5), the agent is suspended (6) and a migration request with the agent's state information is sent to the destination (7). The destination creates a new instance, initializing it with the received state (8), and sends a confirmation to the agent's old host (9), which removes the obsolete instance (10). It then informs the agent's parent and children (11) to update the agent's contact address and resume message transmission towards it. Finally, when the parent confirms the address change (12), the agent is resumed on the new host (13).

Note that the agent binary is "pre-fetched", before contacting the agent's parent and children or suspending the agent. Consequently, the latency of code transfer does not affect the execution of the application. As it will be shown in the next section, this greatly reduces the period during which the application may become unresponsive due to agent migrations performed by the middleware in the background.

4   Performance Measurements

This section presents measurements on the performance of agent creation and migration. The cost of the respective protocols is reported as the number of bytes exchanged through the Network Abstraction component of the middleware, as well as through the (lower-level) ZigBee modem interface; the difference is mainly due to datagram fragmentation. Unless stated otherwise, the network topology is a 4-node chain, with the ZigBee coordinator at the one end acting as the source and other nodes as the destinations of the mobility operations (we report results for up to 3 hops because the network was very unreliable for longer routing paths).

4.1   Agent Creation Overhead

In a first set of experiments, we measure the overhead for creating a leaf agent that requires a special resource, e.g., a user activity sensor. The results for non-leaf agents are similar. The local creation delay for such an agent is 1ms.

Table 1. Cost of agent creation protocol, at the network abstraction layer and ZigBee

agent size [B]	code transport cost [B]		signaling cost [B]		total protocol cost [B]	
	Net Abstr.	ZigBee	Net Abstr.	ZigBee	Net Abstr.	ZigBee
300	352	484	71	107	423	591
600	684	912	71	107	755	1019
900	1032	1392	71	107	1033	1499

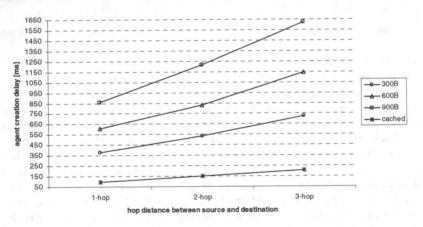

Fig. 4. Agent creation delay as a function of hop distance for different agent sizes

Table 1 analyzes the protocol cost for different agent sizes. The signaling overhead is constant and relatively low, corresponding to one host probe and one agent creation request-reply interaction. Clearly, the dominating component is the code transfer cost, which grows as expected to the agent size. The relative protocol overhead drops as code size increases, but the conversion of datagrams to ZigBee packets costs 35-40%.

Fig. 4 plots the creation time as a function of the hop distance between the source and the destination. The delay rises to the agent size, yet with an economy of scale: creating a 600B agent requires 80% of the time needed to create two 300B agents, and a 900B agent is created in 75% of the time it takes to create three 300B agents. If the agent binary is already cached at the destination, only the signaling cost is incurred, as per Table 1. Hence the respective delay, shown in Fig. 4, is much smaller vs. when code needs to be transferred over the network, yielding an average speedup of 3.7x, 5.8x and 8.4x for a 300B, 600B and 900B agent.

In all cases, the routing overhead is non-negligible. Nevertheless, creating an agent on a remote node directly (as in our middleware) seems a better choice than letting an agent clone itself in a hop-by-hop fashion (as done in other systems). Based on our results, direct creation over 2 and 3 hops is roughly 1.4x and 1.6x faster vs. cloning the agent along these paths.

4.2   Agent Migration Overhead

In a second set of experiments, we measure the migration overhead for a non-leaf agent that is co-located with its parent and has one child on a remote node to which it migrates directly. The runtime state is set to 256B. The delay for performing a corresponding agent suspend-create-init-resume cycle locally is about 2ms.

Table 2. Cost of agent migration protocol, at the network abstraction layer and ZigBee

agent size + state [B]	code transport cost [B]		signaling cost [B]		total protocol cost [B]	
	Net Abstr.	ZigBee	Net Abstr.	ZigBee	Net Abstr.	ZigBee
300+256	352	484	387	543	739	1027
600+256	684	912	387	543	1071	1455
900+256	1032	1392	387	543	1419	1935

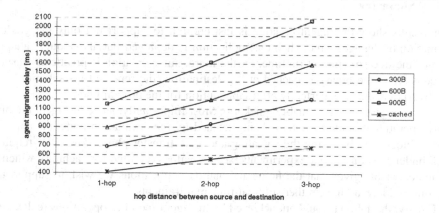

Fig. 5. Agent migration delay as a function of hop distance for different agent sizes

The breakdown of the protocol cost is listed in Table 2. The numbers reported for the code transfer are naturally the same as for agent creation. The signaling cost is much higher though, because it includes the synchronization with the agent's parent and child, as well as the transfer of the agent's state. As a result, the code transfer overhead is less dominant compared to agent creation.

Fig. 5 plots the agent migration time as a function of the hop distance. The trends are similar to the ones observed for agent creation with the respective delays being longer due to the higher signaling overhead. Again, the delay rises with code size, but at a greater economy of scale compared to agent creation, due to the expensive signaling. Namely, the migration of a 600B and 900B agent takes 65% and 57% of the time required to perform two and three migrations of a 300B agent, respectively. For the same reason, the speedup achieved by caching vs. when code transfer occurs is less impressive: 1.7x, 2.2x and 2.9x for a 300B, 600B and 900B agent.

Notably, a direct migration over 2 hops is roughly 1.4x faster vs. two 1-hop migrations, and a direct migration over 3 hops is 1.7x faster than three 1-hop migrations; or 1.5x and 1.8x faster, respectively, when the binary is cached at the destination. This speaks in favor of performing a single long-distance migration vs. several 1-hop ones, as supported by our implementation (the range is set at compile time).

The synchronization with the agent's children also affects the migration delay. To get a feeling of this overhead, we measured the time required to migrate a 600B agent with 256B runtime state while varying the number of its children. In this case, a 5-node star topology is used, with the center node hosting the agent and all children being hosted on different nodes. The recorded delay is 843ms, 874ms, 945ms and 974ms for 1, 2, 3 and 4 children, rising due to the extra signaling required for each additional child. The slight non-linearity from 2 to 3 children is due to the increase in the child information that needs to be transferred along with the agent's runtime state, which happens to exceed the datagram payload limit, requiring an additional datagram to be sent over the network. .

4.3   Summary

Our results show that agent creation is fast enough to support not only the gradual formation of the agent tree when the application is deployed but also a quick adaptation of the tree structure at runtime. Furthermore, since creation is practically instantaneous when the binary is cached at the destination, the repeated instantiation (and removal) of agents is a perfectly affordable option for the programmer.

Agent migration is reasonably quick too. Most importantly, since the agent remains fully operational while its code is being fetched by the destination node, the application is blocked only during the signaling and state transfer phase. The latter requires well under 1 second in our experiments (see the values reported for caching), which is quite acceptable given that the home automation applications we wish to support using our middleware have rather soft real-time requirements.

Finally, the 1-hop throughput achieved by the agent mobility operations (calculated as the number of bytes exchanged through the network abstraction layer in order to perform agent creation/migration divided by the time required to complete this operation), is 12-14Kbps. This is close to the 15Kbps throughput of the Network Abstraction component for reliable datagrams.

5   Application Scenario

In this section we put the benefit and cost of agent mobility in the context of a concrete application scenario. The application structure and logic is kept simple in order to easily follow its operation (the network setup was constrained by the number of nodes at our disposal, as well as the difficulties we encountered in setting up a working network with more than 3 hops). Still, we believe that the results are indicative of the potential gains for more complex applications and larger scale settings.

5.1   Application, Network Topology and Test Scenario

The test application is a subset of the light control application discussed in Sec. 2, namely the part used to infer user absence based on the user activity sensors found in a home. Fig. 6a shows the corresponding tree structure.

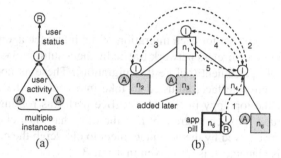

(a)   (b)

Fig. 6. Experiment setup: (a) application tree; (b) nodes, routing topology, and agent placement at different phases of the test scenario.

When a sensing agent does not detect activity, it sends to the inference agent a 1-byte report every 5 seconds. As long as user activity is being detected, the reporting frequency rises to 1 report per 2 seconds. Based on the reports received, the inference agent sends a 1-byte status report to the root every 10 seconds. The size of the root, inference, and user activity sensing agent is 50B, 240B and 24B, respectively.

Fig. 6b shows the network used to deploy and run the application. Unlike n_1, n_4 and n_5, nodes n_2, n_3 and n_6 represent objects which can act as user activity sensors, and therefore can host a user activity sensing agent. The application is launched from n_5 where the root remains fixed. The inference agent can be placed on any node.

The test scenario is as follows:

0. The application is deployed in the network of Fig. 6b without n_3 (added later). The root and the inference agent are created on n_5, while user activity sensing agents are created on n_2 and n_6.
1. Since the message traffic with its children is greater than the message traffic with its parent, the inference agent is migrated on n_4.

2. The agent on n_2 detects user activity and starts reporting at a higher frequency. This increase in message traffic drives the middleware to move the inference agent on n_2.
3. User activity stops, and the sensing agent on n_2 reverts to the normal reporting frequency. Consequently, the inference agent is migrated back on n_4.
4. Node n_3 (which can act as a user activity sensor) is added to the network, leading to the creation of a sensing agent on it. Due to the reporting activity of its new child, the traffic for the inference agent via node n_1 becomes greater than the traffic with n_5 and n_6, so the inference agent is migrated on n_1.
5. Finally, n_3 is removed, the respective user activity sensing agent disappears, and the inference agent is moved back on n_4.

The dashed lines in Fig. 6b denote the migrations that lead to the different placements of the inference agent for each stage.

5.2 Results

Table 3 lists the protocol cost for each migration of the inference agent, as well as the reduction achieved in the wireless network traffic by the resulting placement (after the migration) vs. the old placement (before the migration). These numbers are reported for the ZigBee modem interface, adjusted to take into account the routing cost for each packet as per the topology in Fig. 6b (ZigBee performs routing transparently). The amortization time for each migration, i.e., the time that must elapse in order for the traffic reduction achieved by the new placement to outweigh the cost of the migration that lead to this placement, is also given in Table 3.

Table 3. Cost of migration, wireless traffic reduction achieved by the resulting placement, and the time (of stable operation) required to amortize each migration of the inference agent.

scenario stages	migration of inference agent	migration cost [B]	traffic reduction [B/min]	relative traffic reduction	amortization time [min]
1	$n_5 \to n_4$	873	559	30%	1.5
2	$n_4 \to n_2$	1495	522	22%	2.7
3	$n_2 \to n_4$	769	486	27%	1.6
4	$n_4 \to n_1$	1007	174	8%	5.8
5	$n_1 \to n_4$	511	270	17%	1.9

It can be seen that the migration of the inference agent leads to considerable savings in network traffic, also at a cost that can be recovered in a rather short amount of time. More specifically, the first, the third and the last migration can be amortized in less than 2 minutes, while the second and the fourth migration requires slightly less than 3 and 6 minutes, respectively. Note that when the inference agent returns to a node where it was previously hosted (third and fifth migration), caching reduces the migration cost to 50%, shortening the respective amortization times.

In terms of responsiveness (not shown in Table 3), the delay for creating a user activity agent is about 200ms on average (e.g., the application is deployed in less than half a second). Since user activity agents are created just once on the respective nodes, caching does not apply to this scenario. The average migration delay for the inference agent is 620ms vs. 390ms when the code is cached at the destination. In any case, migration delays are far too insignificant to affect the amortization times or the functionality of the application.

Of course, a migration may turn out to be non-beneficial if the agent tree or the communication pattern between agents changes very fast. In our implementation we use two criteria for identifying and suppressing migrations that are unlikely to be beneficial. Namely, a migration is not performed unless (i) it reduces the amount of network traffic above a threshold and (ii) it can be amortized within a certain amount of time, assuming stable operation. These checks can be done based on information that is locally available. The gains in network traffic that will be achieved after a migration takes place are computed based on the agent's message traffic (the same information is used by the algorithm to decide for a migration), while the migration cost can be estimated using an analytical formula. Both checks are disabled in the experiment. Depending on the thresholds, they would simply lead to fewer migrations.

6   Related Work

Code mobility is supported in many platforms targeted at wireless sensor networks. In the following, we briefly discuss work that is most closely related to ours and give an indicative performance comparison.

Agilla [4] follows a mobile agent approach like POBICOS. However, the application code is written in low-level VM instructions, and the programmer must provide the agent's host discovery and migration logic. Agilla agents communicate indirectly by adding, reading and removing tuples on nodes. Smart Messages [5] (SMs) are mobile code units written in Java, executed using an adapted version of Sun's Java KVM. SMs resemble Agilla agents in that they communicate via the local tag spaces of nodes, and carry their own host discovery and migration code. Also, in both Agilla and SMs, to create an agent/SM instance on a remote node, it must be created locally and then be cloned to the desired destination, typically, in a hop-by-hop fashion. SensorWare [2] allows TCL-based scripts to be injected in a network. Like in Agilla and SMs, the programmer is responsible for providing the logic for cloning/migrating a script, but scripts communicate via message passing. The addressing scheme of SensorWare is very flexible, allowing for attribute-based descriptions combined with the invocation of (default or custom) routing protocols.

MagnetOS [7] statically partitions Java applications, and then places them on different nodes at runtime. Communication transparency is achieved via RPCs. Unless the programmer specifies a placement, the MagnetOS runtime is free to move components between nodes to reduce the network traffic. In DFuse [13], applications are built using so-called fusion points or channels, structured in a hierarchy. Each fusion point takes input from one or more producers and generates output towards one or

more consumers. The initial placement of fusion points, computed off-line, is evaluated at regular intervals to minimize communication and energy consumption, relo-
cating fusion points accordingly.

Agilla is very lightweight, running on MICA2 nodes. All other systems are proto-
typed on PDA devices, while the reported experiments in MagnetOS were done using
laptops. POBICOS seems to be in the middle ground. In fact, given its modest RAM
needs and the fact that it is based on TinyOS, the POBICOS middleware could be
ported on more constrained platforms than the Imote2. The POBICOS VM can also
be implemented efficiently on AVR-compatible microcontrollers, which are a popular
choice for low-end devices.

The differences in the programming abstractions, platform CPUs (Atmel 8-bit microcontroller in Agilla, XScale or StrongARM in other systems, except MagnetOS)
but most notably the wireless technologies used (WLAN in all systems but Agilla,
ZigBee in POBICOS), make a direct performance comparison hard and possibly un-
fair. Still, to give an idea of where our prototype stands, we pick a few cases where a
comparison does not seem entirely out of order. In terms of local operations, creating
a POBICOS agent takes about 1ms vs. 2ms for spawning a SensorWare script, or
2.6ms for the creation of a Smart Message using a single 1KB Java class. The suspend-create-init-resume cycle for a POBICOS agent with 256B of state takes 2ms,
which is the time needed for serializing and de-serializing a Smart Message with a
53B stack frame and 2KB of state. In terms of remote 1-hop operations that do not
involve (significant) code transfer, the creation of a cached POBICOS agent requires
95ms vs. 200ms for weakly cloning a null Agilla agent, 35ms for creating an empty
DFuse channel (over WLAN) and 10ms for spawning a 60B script in SensorWare
(over WLAN). The migration of a cached POBICOS agent with 256B of state that is
co-located with its parent and has one child on a remote node requires 410ms vs.
225ms for a null Agilla agent (in which case no communication endpoints need to be
redirected), 200ms for the relocation of an empty DFuse channel with one producer
and consumer (over WLAN) and 12ms for the migration of a cached Smart Message
with 200B bytes of state (no redirection of communication endpoints, over WLAN).

Overall, given the non-triviality of the underlying protocols and the moderate
throughput of our communication subsystem (e.g., compared to WLAN), the performance of our agent mobility operations is quite satisfactory.

7   Conclusion

We have described the implementation of agent mobility in hierarchically structured
applications, and have presented a performance evaluation of our middleware prototype on Imote2 nodes that communicate over ZigBee. The results show that agent
creation is fast enough to deploy and adapt the application tree structure at runtime.
Furthermore, agent migration can reduce the wireless network traffic significantly,
even for relatively short periods of heavy inter-agent communication. While a faster
communication subsystem would reduce agent mobility delays, the current

performance is already sufficient for a wide range of monitoring and control applications in the home domain.

As future work we plan to implement our own routing on top the native Imote2 radio to experiment with various cross-layer optimizations. We also wish to investigate techniques that will allow the middleware to take smarter agent placement and migration decisions.

Acknowledgements. This work was funded by the 7th Framework Program of the European Community, project POBICOS, FP7-ICT-223984. We also wish to thank Mikko Ala-Louko and Markus Taumberger from VTT in Finland, who implemented the Network Abstraction component of the POBICOS middleware, as well as the entire low-level support for the ZigBee modem.

References

1. Atmel Corporation: 8-bit AVR Instruction Set, rev. 0856H–AVR–07/09 (2009)
2. Boulis, A., Han, C.-C., Shea, R., Srivastava, M.B.: SensorWare: Programming Sensor Networks beyond Code Update and Querying. Pervasive and Mobile Computing Journal 3(4), 386–412 (2007)
3. Domaszewicz, J., Roj, M., Pruszkowski, A., Golanski, M., Kacperski, K.: ROVERS: Pervasive Computing Platform for Heterogeneous Sensor-Actuator Networks. In: Intl. Symposium on a World of Wireless, Mobile and Multimedia Networks (WoWMoM), pp. 615–620 (2006)
4. Fok, C.L., Roman, G.C., Lu, C.: Rapid Development and Flexible Deployment of Adaptive Wireless Sensor Network Applications. In: 25th Intl. Conference on Distributed Computing Systems (ICDCS), pp. 653–662 (2005)
5. Kang, P., Borcea, C., Xu, G., Saxena, A., Kremer, U., Iftode, L.: Smart Messages: A Distributed Computing Platform for Networks of Embedded Systems. The Computer Journal 47(4), 475–494 (2004)
6. Lalis, S., Domaszewicz, J., Pruszkowski, A., Paczesny, T., Ala-Louko, M., Taumberger, M., Georgakoudis, G., Lekkas, K.: Tangible Applications for Regular Objects: An End-User Model for Pervasive Computing at Home. In: 4th Intl. Conference on Mobile Ubiquitous Computing, Systems, Services and Technologies (UBICOMM), pp. 385–390 (2010)
7. Liu, H., Roeder, T., Walsh, K., Barr, R., Sirer, E.G.: Design and Implementation of a Single System Image Operating System for Ad Hoc Networks. In: 3rd Intl. Conference on Mobile Systems, Applications and Services (MOBISYS), pp. 149–162 (2005)
8. Memsic, Imote2 node datasheet,
   http://www.memsic.com/support/documentation/wireless-sensor-networks/category/7-datasheets.html?download=134%3Aimote2
9. Mesarovic, M.D.: Multilevel Systems and Concepts in Process Control. Proceedings of the IEEE 58(1), 111–125 (1970)
10. Pan, M.-S., Fang, H.-W., Liu, Y.-C., Tseng, Y.-C.: Address Assignment and Routing Schemes for Zigbee-based Long-thin Wireless Sensor Networks. In: 67th IEEE Intl. Conference on Vehicular Technology (VTC), pp. 173–177 (2008)
11. POBICOS website, http://www.ict-pobicos.eu

12. Pruszkowski, A., Paczesny, T., Domaszewicz, J.: From C to VM-targeted Executables: Techniques for Heterogeneous Sensor/Actuator Networks. In: 8th IEEE Workshop on Intelligent Solutions in Embedded Systems (WISES), pp. 61–66 (2010)
13. Ramachandran, U., Kumar, R., Wolenetz, M., Cooper, B., Agarwalla, B., Shin, J., Hutto, P., Paul, A.: Dynamic Data Fusion for Future Sensor Networks. ACM Transactions on Sensor Networks 2(3), 404–443 (2006)
14. Texas Instruments: Z-Accell Demonstration Kit, http://focus.ti.com/docs/toolsw/folders/print/ez430-rf2480.html
15. Tziritas, N., Loukopoulos, T., Lalis, S., Lampsas, P.: On Deploying Tree Structured Agent Applications in Networked Embedded Systems. In: D'Ambra, P., Guarracino, M., Talia, D. (eds.) Euro-Par 2010. LNCS, vol. 6272, pp. 490–502. Springer, Heidelberg (2010)

Vital Responder – Wearable Sensing Challenges in Uncontrolled Critical Environments

Miguel Coimbra[1] and João Paulo Silva Cunha[2]

[1] Instituto de Telecomunicações, Faculdade de Ciências da Universidade do Porto
mcoimbra@fc.up.pt
[2] IEETA/Universidade de Aveiro and DEEC, Fac. Engenharia da Universidade do Porto
jcunha@ieee.org

Abstract. The goal of the Vital Responder research project is to explore the synergies between innovative wearable technologies, scattered sensor networks, intelligent building technology and precise localization services to provide secure, reliable and effective first-response systems in critical emergency scenarios. Critical events, such as natural disaster or other large-scale emergency, induce fatigue and stress in first responders, such as fire fighters, policemen and paramedics. There are distinct fatigue and stress factors (and even pathologies) that were identified among these professionals. Nevertheless, previous work has uncovered a lack of real-time monitoring and decision technologies that can lead to in-depth understanding of the physiological stress processes and to the development of adequate response mechanisms. Our "silver bullet" to address these challenges is a suite of non-intrusive wearable technologies, as inconspicuous as a t-shirt, capable of gathering relevant information about the individual and disseminating this information through a wireless sensor network. In this paper we will describe the objectives, activities and results of the first two years of the Vital Responder project, depicting how it is possible to address wearable sensing challenges even in very uncontrolled environments.

Keywords: Wearable sensing, sensor networks, biomedical signal processing.

1 Introduction

A recent study on firefighters in the United States [1] showed that 45% of the deaths that occur among U.S. firefighters, while they are on duty, are caused by heart disease. This number is twice as high as for police officers and three times as high as the average incidence of heart disease at work. Furthermore the study shows that the risk of death from coronary heart disease during fire suppression is approximately 10 to 100 times as high as that for nonemergency events. These facts clearly show that for a firefighter the most life threatening condition besides factors such as direct contact to fire or chemicals, is the condition of his heart. Factors that have an obvious high impact on the cardiovascular system are stress and fatigue, which might also be triggering factors for its overload.

F. Martins, L. Lopes, and H. Paulino (Eds.): S-Cube 2012, LNICST 102, pp. 45–62, 2012.

The Vital Responder project is an interdisciplinary research project formed by teams from Institute of Electronics and Telematics Engineering of Aveiro (IEETA), Carnegie Mellon University (CMU), Instituto de Telecomunicações (IT) in Porto and Aveiro, and BioDevices, S.A. The goal of the Vital Responder research project is to explore the synergies between innovative wearable technologies, scattered sensor networks, intelligent building technology and precise localization services to provide secure, reliable and effective first-response systems in critical emergency scenarios. For this goal the estimation of stress and fatigue in first responders is a main concern to prevent cardiac failure and is addressed by using a wearable technology (Fig.1) to obtain information on the firefighter's cardiovascular status via electrocardiography (ECG).

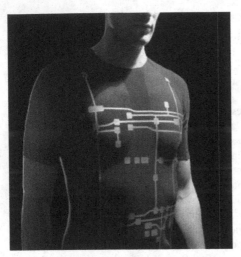

Fig. 1. The Vital Jacket is a wearable vital sign monitoring device. Due to its design in form of a light T-shirt it allows to record one lead clinical quality ECG with a sampling frequency of 200 Hz without restricting the freedom to move.

As in other areas of telemedicine, vital signs monitors are an important part of medical equipment market that has been moving from hospitals to the patient's home [2]. In critical situations, the use of these portable vital signs monitors has proved to be valuable [3]. In this domain, intelligent wearable garments are emerging as the most promising technology [4][5][6]. In a prevous study [7] Cunha et al. showed that the market of wearable vital signs monitors presented eight relevant players. We have evaluated their "integration of vital signs" and "user mobility" characteristics, and results showed two main players: Sensatex's "Smart Shirt", and Vivometrics' "LifeShirt". These are but rare examples of wearable technology for vital signs monitoring, and Vital Jacket (Fig.1) is clearly one of the most successful products, being a comfortable high clinical quality medical device compliant with the EU MDD directive 42/93/CE and manufactured through a ISO9001 and ISO13485 certified process. This device is certified for clinical usage in more than 30 countries. Regarding the estimation of stress in real environments using wearable sensors, recent research has been mostly focused on driving scenarios [8,9] or to estimate stress during high performance jet flights [10]. Some interesting results were obtained by analyzing the sympathovagal balance between sympathetic and parasympathetic activity using heart rate variability (HRV) measurements [11].

It is thus quite clear that the Vital Responder project, when compared to current scientific literature, addresses the very demanding challenges of a highly uncontrolled environment, such as firefighters in action. Will our sensors gather all the required signals with enough quality? Can we transmit this information robustly? Is it possible to estimate stress using conventional HRV measures when there is intense physical activity too? In this paper we will describe the successful experience of the Vital Responder project, including not only a generic presentation on its structure (Section 2), as well as some specific topics associated with the difficulties of attempting to monitor vital signs in uncontrolled environments. We will discuss how standard technologies needed to be adapted to forest fire environments (Section 3), how a sophisticated data annotation system had to be designed for contextualizing the gathered vital signs (Section 4) and how current state-of-the-art biomedical signal research results are not easily transferred to uncontrolled environments (Section 5).

2   The Vital Responder Project

In order to address the demanding challenge of monitoring vital signs of first responders in action, a multi-disciplinary approach was required, resulting in the project structure depicted in Fig.2. This includes designing novel wearable technologies (Task 2), robust wireless signal transmission (Task 3), middleware integration (Task 4) and data annotation and analysis (Task 5).

The Vital Responder project is an international cooperation between Portuguese and North American partners funded by Fundação para a Ciência e Tecnologia (CMU-PT/CPS/0046/2008). Its budget totals 519.456€ and it will end in September 2012 after a 3 year duration. Besides academia and industry partners, the project includes two end-user firefighter corporations, namely Bombeiros Voluntários de Amarante and Bombeiros Sapadores de Vila Nova de Gaia, both in Portugal. Some relevant achievements after only two years include 7 MSc thesis (4 finished, 3 ongoing), 7 PhD thesis (1 submitted, 6 ongoing), 4 journal papers (1 published, 3 submitted), 15 conference papers published, more than 25 public presentations, over 1300 hours of recorded vital signs, 10 new VR Vital Jackets deployed and 2 patents (1 accepted, 1 submitted).

More information about the Vital Responder project can be obtained in the following website: www.vitalresponder.org.

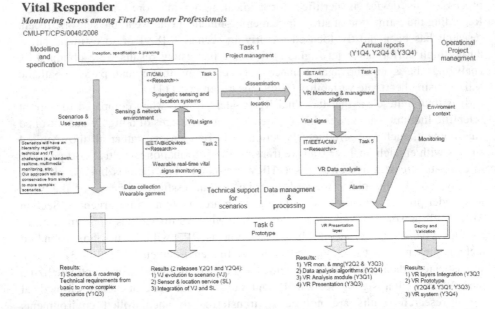

Fig. 2. Diagram of the structure of the Vital Responder project, highlighting its multidisciplinary nature with tasks ranging from sensor technology, networks, signal processing, machine learning and human-computer interaction.

3   Wearable Sensing

The main objective of Vital Responder regarding wearable sensing is to devise the next generation of wearable intelligent garments for vital signs and other body information suitable for the demanding requirements of first responder professionals. This includes not only the sensing equipment but also its multimedia integration that can support the transmission, recording, annotation, processing and delivery to the final user. Although the Vital Responder project addresses many other challenges (ad-hoc networks for forest fires, building evacuation systems, indoor location systems, arrhythmia detection, fatigue estimation using accelerometers, etc.), we will focus on three specific ones, given their relevance for uncontrolled environments:

-    The new **VR Vital Jacket wearable shirt**, adequate for the specific requirements of a firefighter's working conditions. (Section 3.1)
-    Novel sensing technologies besides Vital Jacket including, a **real-time blood pressure estimation system,** a **helmet that can record CO, barometric pressure and temperature**, and a new **dry electrode** that is adequate for high temperatures. (Section 3.2)
-    The **iVital multimedia integrator** that can be used by a firefighter team leader for viewing the position and vital signs of his team members in real time in an outdoor environment. (Section 3.3)

3.1 Vital Jacket for First Responders

According to the needs identified during our user studies of firefighters in action, a new generation of Vital Jackets has already been produced, tested and deployed in the field. The main problem we had to solve was that our previous process used textile fabrics with a mixture of elastane (28%) and polyamide (62%) and elastane is heat sensitive and may burn the skin of its wearer. Due to this fact, and following international regulations, the firefighters' clothing had to be made with less than 2% of elastane. This means we had to "re-invent" the way we were embedding micro-cabling and micro-electronic components in the shirt to comply with this requirement.

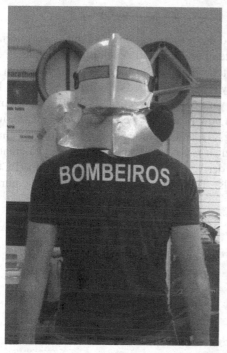

Fig. 3. The new VR Vital Jacket for firefighters had to comply with both international safety regulations for textiles under high heat conditions and the Portuguese law on firefighter garments. As a result, the new shirt (left) visually mimics the normal shirt worn by firefighters and now has less than 2% elastane. Temperature, barometric pressure and carbon monoxide sensors were also embedded in a firefighter helmet model that complies with all EU safety regulations (right).

This development was performed by Biodevices in cooperation with Petratex and IEETA, involving a task-force of textile engineers, micro-electronics experts and textile machine technicians. This highly multi-partner (university and industry) interactive and iterative process focused on devising and implementing the new ways to embed micro-electronics/micro-cabling into non-elastane textile fabric. Furthermore, we had also to design the new VR Vital Jacket version (Fig. 3) following all the standards imposed by international and Portuguese law on a firefighter's garment. Several options were tested that included polyamides, high performance polyamides (HPPA) and cotton. The final solution reduced the elastane from 28% to 2% and incorporated 98% cotton instead of polyamides fabric. Moreover, we could design the new garment in a way that all the outer part (that heats more than the inner part) only had cotton fabric, making it even more protective and heat proof, being at the same time highly comfortable for a firefighters' daily routine. Changes in the fabrication process that involves nosew® textile technology and new micro-cabling components were developed after a long trial and error procedure.

3.2   Novel Sensing Technologies

During Vital Responder we have addressed the possibility of continuously monitoring blood pressure, measuring CO levels and temperatures and developing a dry electrode (without gel). Blood pressure is an important vital sign in the stress study, known to be more sensitive than heart rate variability (HRV) extracted from ECG. Motivated by this, we have initiated during the second year of the project the RTABP (Real Time Arterial Blood Pressure) project. This is a new prototype wearable device that allows comfortable monitoring of subjects (Fig. 4) and includes the estimation of their arterial blood pressure in an online manner. It is based on an algorithm that calculates arterial pulse-wave transit time (PWTT) derived from the electrocardiogram (ECG) and photoplethysmography (PPG) signals. The physiological signals are acquired

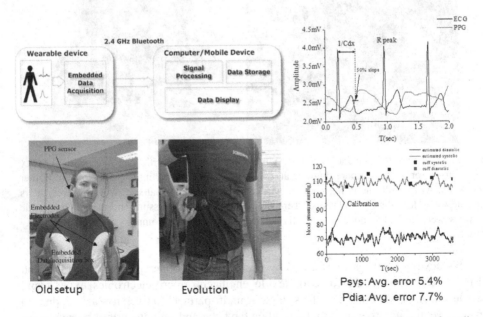

Fig. 4. The Real Time Arterial Blood Pressure device. By increasing the number of cardiac parameters that can be sensed in the field we will pave the way for superior cardiac health assessment in the future.

synchronously in real-time through a textile embedded and a custom built signal acquisition unit (extension of the Vital Jacket system). The sensor used for photoplethysmography (ear probe) permits the quantification of oxygen saturation levels, which is an important indicator of a first responder's physical condition. It also allows the storage of all data for off-line analysis or to transmit the data wirelessly to a computer or mobile device.

Other variables that were identified for monitoring were Carbon Monoxide (CO), barometric pressure and the temperatures that a firefighter was exposed to while on duty. To approach this we have developed the FREMU (First Responder External Measurement Unit). This unit is a device that is inserted into the fire-fighter's helmet that allows for continuous monitoring of environmental conditions in hazardous situations (Fig. 3). This Unit includes a Carbon Monoxide (CO) sensor, temperature sensor and a Barometric Pressure sensor. The CO sensor is important to assess local environmental conditions and give real-time feedback and alarms to the team leaders and commanders. Furthermore we can estimate altitude from values from the barometric pressure sensor. This sensor will be an important tool in order to locate first responder personnel in the field. Communication is handled wirelessly via Bluetooth and the power is drawn from a battery. The unit can be attached to the equipment (e.g. helmet) of the fire-fighter. This device is on lab tests and will be deployed in the next months.

Finally, we have noticed that the presence of high temperatures dries out the liquid gel used in Ag/AgCl ECG electrodes (most commonly used) and the signal loses quality faster than in normal conditions with room temperature. This poses a new problem to our firefighter's scenarios. So, we revisited previous R&D of the IEETA group in the area of active electrodes and developed a novel dry active ECG electrode that does not need gel to pick-up biosignals, being more resilient to temperature. The solution takes advantage of a hybrid organic-inorganic material and embedded microelectronics to perform impedance buffering. This was a major development that has been published on IEEE Sensors journal and resulted in the submission of a patent.

3.3   Multimedia Integration

The iVital system (Fig.5) is a solution to monitor teams of first responders, supporting the role of a team coordinator, with access to the aggregate data. This data includes individual vital signs (e.g. ECG) and location (through GPS, when available). iVital is able to trigger alerts to the end user that could be crucial to support decision in emergency operations, such as forest fires or rescue missions. The system has three main components: a wearable vital signs data collecting unit, using the Vital Jacket; a mobile device for processing and relay (DroidJacket running on a Android smartphone) and a mobile team coordination station (the iVital Base Station running on a iPad tablet) (Fig.6). DroidJacket is the main processing and relay element of the iVital system. It is responsible for processing over the received data in order to identify specific situations from technical issues (e.g. loss of connectivity) to more critical events found in ECG (such as arrhythmias) or in activity patterns (fall or low activity events) by means of the data received by the accelerometers and the smartphone gyroscope. The Base Station handles the incoming data from the DroidJackets through a Wi-Fi connection and displays the location of the team members, as well the individual status of both vital signs (e.g. real time heart rate, heart rate history, ECG) and the mobile device (e.g. battery and connection status).

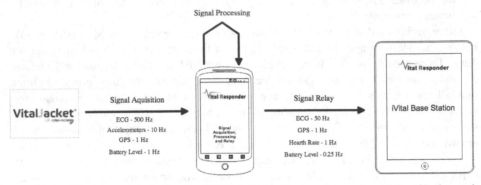

Fig. 5. The *iVital system*, including the *VR Vital Jacket*, the smartphone-based *DroidJacket*, and the tablet-based *iVital Base Station*

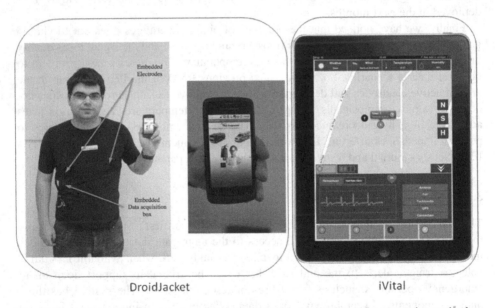

DroidJacket   iVital

Fig. 6. Two of the Vital Responder prototypes. DroidJacket is the main processing and relay element of the iVital system. It is responsible for processing over the received data in order to identify specific situations from technical issues to more critical events found in ECG. The iPad Base Station handles the incoming data from the DroidJackets through a Wi-Fi connection and displays the location of the team members, as well the individual status of both vital and the mobile device.

4   Signal Annotation for Uncontrolled Critical Environments

Gathering large amounts of vital signals from first responders in critical events is but the first step for researching automatic stress and fatigue estimation algorithms. The currently used Vital Jacket version for firefighters (FF) is capable of gathering clinical quality electrocardiogram, 3 axis accelerometers and GPS signals that are sent to a base station, in real time, in order to be analyzed. A new version is on the way that will add temperature, barometric pressure and carbon monoxide through the helmet embedded unit. However, signal processing and machine learning research, aiming to combine these signals into an estimation of an individual's stress levels, requires the data to be complemented by adequate annotation that can contextualize it. Did an individual's heart rate rise because he was stressed? Or did he simply start running? Was he doing a training exercise or in a real dangerous life-threatening situation? For this purpose, we have created the Vital Analysis Framework (VA), which is a smartphone based solution capable of annotating physiological signals of FF in action with both context (details about the event the FF was involved) and perceived psychological stress levels (retrieved from the analysis of psychological questionnaires).

4.1   Annotating Stress Levels

When investigating stress, it seems important to acknowledge that over the last decades the term has increased popularity across different areas of study such as behavior and health sciences. As a result, it remains difficult to define the concept, at least in simple terms [12], and therefore the necessity to use standardized measures of assessment. We will follow the transactional model of stress, defined as "a situation that taxes or exceeds one's personal resources or threatens the person well-being has the potential to cause stress" [13] (p.19). Thus, the emotions experienced and physiological responses are initiated due to the person interpretation of the event and its perceived meaning to their well-being [14].

Following these theoretical conceptualization of stress, its assessment should combine physiologic and psychological measures [15], including a longitudinal research design and decreased time delay between real world experience of an event and stress ratings [13] since previous research methodologies rely mainly on self-report measures of the concept (e.g., questionnaires) and are fulfilled in paper several hours after the event [16]. To overcome issues related with recall errors, the current study will use the experience sampling method (ESM) developed for in situ recording [17]. ESM in the current study will ask firefighters to rate their stress levels at predetermined times (e.g. beginning, and end of the day and following an event). While ESM originally relied on paper surveys, for the purpose of the current study, ESM will be conducted using a smartphone based framework as successfully used in previous research [18].

4.2   The Vital Analysis Prototype

Given that our target is firefighters in action during real events, we designed a simple to use smartphone framework that can adapt itself to their daily routines. Three annotation methodologies were designed that together provide context to an event and to the collected signals (screens depicted in Fig. 7). These are the Stress Annotation Methodology (stress levels annotation), the Event-driven Annotation Methodology (event context), and the Voice Annotation Methodology ("break the glass" mechanism).

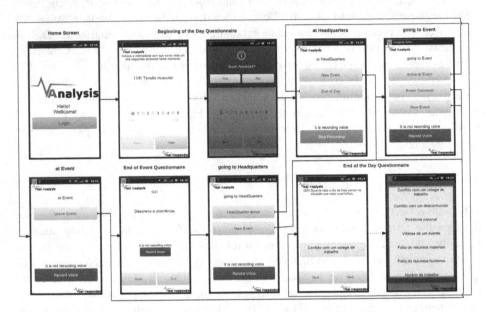

Fig. 7. System image of the Vital Analysis prototype

Stress Annotation Methodology - Due to the complexity of the definition of the stress concept, several self-report measures exist. Despite differences in types of questions, those vary according to the context and population under study. Most self-report measures aiming to access stress levels include questions related with physical and cognitive symptoms of stress. Thus, following this principle, our measures of stress included 4 questions related with physical and 4 questions related with cognitive aspects, used previously in validated stress questionnaires. An example of a physical symptom question is: "I have a stiff neck"; an example of a cognitive symptom question is "I lack concentration". Participants were asked to rate each item on a free scale ranging from "0" to "4" kinds of ratings, where a rating of "0" was used to represent not felt at all, and a rating of "4" extremely felt. These questions were fulfilled at the beginning and end of the day, aiming to evaluate whether there were alterations in stress symptoms experienced, from beginning to end of the day. Furthermore, end of the day and beginning of the day stress symptoms mean scores will be subtracted in order to accomplish an overall mean score, symbolizing accumulated stress symptoms over the day. Internal consistency of the 8 questions was calculated using Cronbach's alphas. This value provides a coefficient of reliability, and it is used as a measure of internal consistency for participants' answers. In the current study, the Cronbach's alphas found for the 8 questions was 0.93. Additionally, another question was fulfilled after each event, indicating stress appraisal of the event. Participants were asked to rate how they appraised each stressful event, on a free scale ranging from "0" to "4" kinds of ratings, where a rating of "0" was not at all stressful, and a rating of "4"was extremely stressful.

Event-driven Annotation Methodology - The Event-driven Annotation gives us the possibility to detail an event by dividing it into several predefined stages, allowing us to evaluate and quantify the collected physiological signals differently for each one. These predefined stages are the basic stages for every single event, and are usually consecutive. A normal event starts with an emergency call, followed by the trip to the event location, the event itself, and finally the return trip to the headquarters. Nevertheless special occasions can occur such as: a high priority call during any period of other event; or the cancellation of an event. This workflow is represented in Fig. 7, where all the options available in the framework can be seen.

Voice Annotation Methodology - Motivated by the unpredictability of a firefighter's job we have design the Voice Annotation Methodology. This methodology will be our "break the glass" mechanism, allowing at least one of the firefighters to report unexpected activities that happen during an event, or add valuable psychological information, allowing for rich and expressive contributions. This methodology is also used to allow the user to add annotation to outside events and to enrich the data gathered using the previous methodologies.

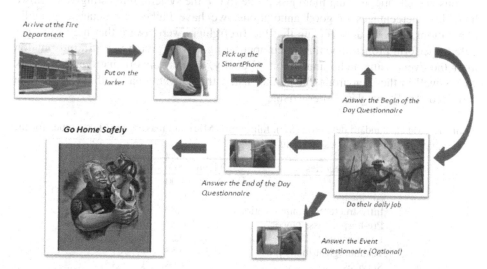

Fig. 8. Integration of the Vital Jacket device and the Vital Analysis prototype in a firefighter's daily routine. More than 4600 hours of Vital Analysis were collected with this methodology.

4.3   Results

The dataset compiled for this study was collected from 12 firefighters, with a mean age of 37.8 and a standard deviation of 5.3, between July, 2011 and January, 2012. During this time we have collected more than 4600 hours of Vital Analysis data from which we have retrieved a total of 717 answers to the event questionnaires from 454 different events. We have also retrieved 319 stress evaluations from the differences between the beginning of day and end of day questionnaires.

To evaluate the usability of our framework we have collected the official information about the events, already gathered using today's firefighter (FF) protocol, and we compared it with our annotations. The measures chosen were: the percentage of real events that were annotated; the percentage of annotations that were done correctly, in which a correct (good) annotation is one that has all the stages implemented in the Event-driven Methodology (in subsection III-B), with a time difference between them superior to 1 minute; the percentage of annotated events with audio annotation; and the percentage of event questionnaires with audio annotation. Results showed that our framework was used in 53.5% of all events and that 64.2% of them were correctly annotated. We can generically consider this as a good result, given the harsh environments that these FF have faced. Nevertheless, some results require a special attention. Low percentages of annotation are obtained when a single FF is sent to an event with VA, which was an expected result. Our solution requires that the FF by himself both remembers and has the time to use the VA which does not exploit the redundancy of the team. Another interesting result is the low percentage of good annotations in events where we have 3 or 4 firefighters with VA present. We would expect to see an increase in this percentage but an explanation might be that situations when many men are deployed tend to be more serious and chaotic, making them less prone to use the system. Interestingly, when we have low percentages of good annotations, we have higher percentages of audio annotations, making us speculate that the firefighters were aware that they were not able to perform proper annotations, compensating this by giving us extra information after the event using audio annotation. Globally, results support that our framework works well in these environments, either using the conventional methodology or the provided "break the glass" alternative.

Table 1. Mean, standard deviation (SD), minimum (Min) and maximum (Max) values for the stress appraisal of various types of events

Event categories	Mean	SD	Min	Max
Fire	1,23	1,00	0	4
Accident	1,40	0,86	0	3
Infrastructure/Communications	0,50	0,80	0	2
Pre-hospital assistance	1,11	0,81	0	4
Legal conflict	0,80	0,83	0	2
Technological/Industrial	0,50	0,55	0	1
Services	1,01	1,03	0	3
Activities	0,77	0,73	0	2
Total	1,10	0,92	0	4

Table 1 provides the mean and standard deviation for the stress appraisal of various events. To analyze whether means for each stress category differed, One-Way Anova analysis was conducted. As expected, we found significant differences between the stress appraisal categorization of the various events, as can be observed by F value (F= 2.518) which is found by dividing the between group variance by the within group variance. When testing the statistical significance, the p-value found was 0.01. It is important to highlight, that higher ratings of stress were provided for certain events such as fire, accidents and pre-hospital assistance. These findings are similar to those found in the literature using detailed psychological methods to assess stress among firefighters. This fact is likely to support the accuracy of the measure to assess stress appraisal of FF across different events.

5   Associating ECG features with Stressful Activities

Most interesting results on stress estimation using ECG signals are based on heart rate variability (HRV) measurements [2,3,4,5]. We are however concerned that in these scenarios the individual being monitored is not under significant amounts of fatigue or short-term physical stress, which is not the case of first responders in action. We aimed to inspect how these cardiac features obtained from ECG signals behaved in a scenario where both high levels of fatigue and stress are expected. To accomplish this we measured the associations between HRV features and certain types of emergency and non-emergency events in which varying degrees of stress are expected.

5.1   Materials

The data used for these experiments consist of records from five male firefighters from the Bombeiros Voluntários de Amarante, which is a team of volunteer firefighters based in Amarante, Portugal. To collect the data, in the beginning of the working day the firefighters put on a shirt under their clothes with the ability to collect clinically valid ECG signals during very long periods (VJ - Vital Jacket) [19]. The VJ then records the firefighter's ECG signal together with a time stamp until the end of his shift, which in most of our recorded cases takes 8 to 12 hours. Afterwards the ECG signal is transferred from the internal memory of the Vital Jacket to a database.

Together with the ECG data recorded by the Vital Jacket the timestamps and types of activities occurred during a specific shift were logged. We have used the official daily log of each firefighter that participated in this study. The Portuguese law implies that these logs include the following information: date and time of the beginning and of the end of every event; type of event, according to a national scheme for classifying events; and the tactical position within the team. The categorization of events is divided into nine main classes with a varying number of sub classes. The main classes are: Fire, Car Accident, Infrastructures/ Communications, Pre-Hospital Assistance, Legal Conflict, Technological/ Industrial, Services, Activities and Civil Protection Events.

With this setup around 447 hours of ECG were recorded between February 2010 and July 2010 from five firefighters. The average age for these firefighters was 35.4 years, with a maximum of 41 years and a minimum of 24 years and at least five years of experience in firefighting. The harsh environment and unexpected situations that these individuals face led to a substantial amount of bad signal recordings due to inadequate electrodes, broken hardware, and incorrect time stamping of the data. As a result, this study led to a clear improvement of the data collection system, which is now much more robust and reliable. After this screening, 238 hours of recordings were selected, out of which 59 hours were collected during missions. The distribution of these 59 hours divided by types of events can be seen in Fig. 9.

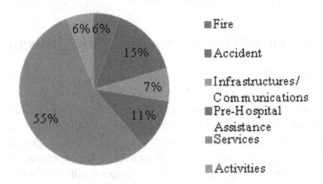

Fig. 9. Distribution of the 59 hours of collected ECG signal according to the type of event

5.2   HRV Features

The detection of QRS complexes used is based on the algorithm by Pan and Tompkins [20] implemented together with further improvements in the open source EP Limited QRS detection software [21] which was used to detect R peaks in the ECG recordings. Besides instantaneous heart rate (HR) measured in beats per minute (bpm), six standard measurements proposed by the task force of the European Society of Cardiology and the North American Society of Pacing and Electrophysiology were used for the heart rate variability (HRV) analysis. The following three time domain measurements were used:

- **SDNN:** Standard deviation of all NN (normal to normal beat) intervals in ms;
- **RMSSD:** Square root of the mean of the sum of the squares of successive differences between NN intervals in ms;
- **HRV triangular index (HRVti):** Total number of all NN intervals divided by the height of the histogram of all NN intervals measured on a discrete scale with bins of 7.8125 ms.

Together with the time domain measures, three frequency domain measures were also used. To obtain these measures the series of NN intervals was transformed to a power spectral density (PSD) using the Lomb Periodogram [22]. The spectrum then was analyzed using:

- **Low frequency (LF) part**: ranging from 0.04-0.15 Hz;
- **High frequency (HF) part**: from 0.15-0.4 Hz.

The two previous measures were then divided by the total frequency in the 0-0.4 Hz frequency band. Finally, the third measure is:

- **LF/HF**: ratio between the LF and HF component.

All measures were calculated in consecutive windows of five minutes, until the whole record was covered.

5.3 Associating ECG Features with Types of Events

Association measures are important and useful when evaluating a predictive relation between two variables [23]. The most used measures are the correlation measures, which are adequate for continuous variables, e.g. Pearson correlation. However, in the presence of discrete variables (such as the stress ranking) these popular measures could not be applied. In this special case, another kind of association measures should be used. The ideal measure, in our study, should describe the stress/fatigue as a monotonically nondecreasing mathematical function of the HRV measurements. The Kim's dyx measure is described as adequate for the present problem [23]. However, the Pk measurement by Smith et al. [24] was used, which is an easier to interpret modification of the underlying dyx measure by Kim [25] and often used to evaluate the quality of indicators of anesthetic depth. Briefly, when comparing indicator values (in this case the HRV measurements) to an ordinary scale (the stress ranking) the value of Pk with a range from 0 to 1 can be interpreted as the probability of a concordant relationship of both sides, which means that if the indicator value increases, the assigned level of the ordinary scale is also increasing. The full methodology is depicted in Fig. 10.

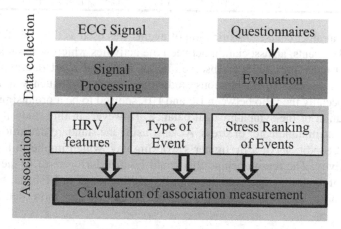

Fig. 10. Data processing methodology

5.4   Results

Due to the answers given in the questionnaires the categories were ranked from low to high stress as following: 1) services, 2) activities, 3) pre-hospital assistance, 4) infrastructures / communications, 5) fire and 6) car accidents. This ranking method produced the same ordering for stress and for fatigue.

Calculating the Pk measure for all six classes and the mentioned HRV measurements including heart rate resulted in Pk values around 0.5 which means a probability of 50% to predict an increased stress level according to an increased value of the measure. But considering only the extreme levels of low stress and high stress at both ends of the scale showed interesting results, as observed in Table 2. The mean values over all five minute segments for the five firefighters are shown divided into the three extreme types of activities: services, fires and car accidents.

Although this is a rather generalized approach, it can be seen that the average heart rate, as the strongest predictor, clearly distinguishes differences between these types of activities and in 76% of all cases supports the order estimated by the questionnaires. Also, the time domain features, mainly HRVti and SDNN, show a negative association with our ordering. This means that the HRV tends to go down during events, which were ranked as more stressful, like fires and car accidents, compared to a low stress service activity, like cleaning route, unblocking passageways or patient transportation.

Table 2. Pk association measure and mean values for 5 minute segments over five firefighters according to three different activities: services, fires and car accidents

Ranking position	Classes	HR (bpm)	SDNN (ms)	RMSSD (ms)	HRVti	LF norm	HF norm	LF/HF	rand
1	Service	87.07	77.44	44.69	12.79	0.35	0.11	5.56	
5	Fire	101.54	65.12	31.01	10.21	0.34	0.09	4.71	
6	Car ac.	103.96	57.71	30.63	9.47	0.41	0.10	5.30	
P_k		0.76	0.39	0.43	0.32	0.57	0.49	0.54	0.53

Although a rather coarse classification of events was used as the basis for the stress ranking of the events, an association between the activities which were ranked as least stressful and most stressful was observed as an increased heart rate and decreased HRV in the time domain. In previous studies the standard frequency domain HRV features, especially the ratio between LF and HF, showed to be a promising parameter as an estimation of the sympathovagal balance. However, in this study it turned out not to be a potential indicator of stress. At least using the standard definition under the uncontrolled conditions of a firefighter's working day, which implies a high level of physical activity. The reason behind this could be that during most of the events the largest part of the power spectrum does not lie within the HF or LF band, but in lower frequency parts. These parts probably should be assessed in more detail in future works.

6   Discussion

Wearable sensing will most probably play a major role in all future health monitoring. The "self monitoring" or the "quantified self" concepts are new and promising trends that will take advantage of this type of technologies. Novel sensing devices appear every year and smartphones are now a very powerful and popular tool that can be exploited. In this paper we have shared some of the experiences of the Vital Responder project, with a stronger emphasis on the problems that arise by demanding uncontrolled critical environments. These are most probably some of the most demanding environments for approaching vital signs monitoring. Sensing technology must adapt and cope with these environments, gathering large annotated datasets that can be useful for signal processing and machine learning research is not a trivial challenge, and traditional state-of-the-art biomedical signal processing research does not necessarily translate well to uncontrolled environments. These are but some the various take home messages produced by the Vital Responder projects and more details can be obtained in the project website (www.vitalresponder.org) and the many scientific publications that resulted from it. We look forward to continue pursuing this line of research in the coming years.

Acknowledgments. The authors would like to thank the Cmdt. Jorge Rocha and the firefighters of Bombeiros Voluntários Amarante, as well as Cmdt. Salvador Almeida and the firefighters of Companhia Bombeiros Sapadores de Gaia for their participation in this study. Furthermore, we would like to acknowledge all the researchers of the different multidisciplinary teams involved in the Vital Responder project, leaded by the co-PIs: Priya Narasimhan from CMU, João Barros from IT/FEUP, Luis Meireles from Biodevices SA, Verónica Orvalho from IT/FCUP and Nuno Borges de Carvalho from IT/UAveiro. This work could not have been performed without their active contribution. Finally, we would like to thank both Pedro Gomes and Johannes Pallauf, which were the main authors of the papers in which Sections 4 and 5 were based. The present work was partly supported by FEDER grant CMU-PT/CPS/0046/2008.

References

1. Kales, S.N., Soteriades, E.S., Chistophi, C.A., Christiani, D.C.: Emergency duties and deaths from heart disease among firefighters in the United States. N Engl. J. Med. 356, 1207–1215 (2007)
2. Stefanov, D.H., Bien, Z., et al.: The Smart House for Older Persons and Persons with Physical Disabilities: Structure, Technology, Arrangements and Perspectives. IEEE Transactions on Neural Systems and Rehabilitation Engineering 12(2), 228–250 (2004)
3. Gao, T., Massey, T., et al.: The advanced health and disaster aid network: A light-weight wireless medical system for triage. IEEE Transactions on Biomedical Circuits and Systems 1(3), 203–216 (2007)

4. Lymberis, A., Olsson, S.: Intelligent Biomedical Clothing for Personal Health and Disease Management: State of the Art and Future Vision. Telemedicine Journal and e-Health 8(4), 379–386 (2003)
5. Lymberis, A., Rossi, D.: Wearable e-health systems for personalised health management: State of the art and future challenges. IOS Press, Amsterdam (2004)
6. Huang, H.-P., Hsu, L.-P.: Development of a wearable biomedical health-care system. In: IEEE/RSJ International Conference on Intelligent Robots and Systems, IROS 2005 (2005)
7. Cunha, J., Cruz, I., et al.: SIMoD: Product Strategy Study. SIMoD Project Report, University of Aveiro (2005) (in Portuguese)
8. Healey, J.A., Picard, R.W.: Detecting stress during real-world driving tasks using physiological sensors. IEEE Trans. Intell. Transp. Systems 6, 156–166 (2005)
9. Lee, H.B., Kim, J.S., Kim, Y.S., Baek, H.J., Ryu, M.S., Park, K.S.: The relationship between HRV parameters and stressful driving situation in the real road. In: Proc. 6th Int. Special Topic Conf. Information Tech. Applications in Biomedicine, pp. 198–200 (2007)
10. Di Rienzo, M., et al.: Assessment of gravitational stress on heart rate variability during meneuvers on high performance jet flights. In: IEEE Proc. Annual Int. Eng. in Med. and Biol. Soc., pp. 3457–3459 (2010)
11. Malik, M., et al.: Heart Rate Variability: Standards of Measurement, Physiological Interpretation, and Clinical Use. European Heart Journal 17, 354–381 (1996)
12. Levine, S.: Handbook of Stress and the Brain, ch. Stress: An historical perspective, pp. 3–23. Elsevier (2005)
13. Lazarus, R.S., Folkman, S.: Stress, Appraisal, and Coping. Springer (1984)
14. Lovallo, W.R.: Stress & health: Biological and psychological interactions. Sage Publications (2005)
15. Semmer, N., Grebner, S., Elfering, A.: Beyond Self-report: using observational, physiologival, and situation-based measures in research on ocuppational stress. In: Research in Occupational Stress and Well-being, vol. 3, pp. 205–263. Emerald Group Publishing (2003)
16. Cohen, S., Kessler, R., Gordon, L.: Measuring stress: a guide for health and social scientist. Oxford University Press (1997)
17. Hektner, J., Csikszentmihalyi, M.: The experience sampling method: Measuring the context and content of lives. In: Handbook of Environmental Psychology. John Wiley Sons, Inc. (2002)
18. Klasnja, P., et al.: Using wearable sensors and real time inference to understand human recall of routine activities. In: UBIComp 2008 (2008)
19. Cunha, J.P.S., et al.: Vital Jacket: A wearable wireless vital signs monitor for patients' mobility in Cardiology and Sports. Presented at the 4th International ICST Conference on Pervasive Computing Technologies for Healthcare 2010, Munich, Germany (2010)
20. Pan, J., Tompkins, W.J.: A real-time QRS detection algorithm. IEEE Trans. Biomed. Eng. 32, 230–236 (1985)
21. Hamilton, P.: Open source ECG analysis. In: Proc. Computers in Cardiology, pp. 101–104 (2002)
22. Lomb, N.R.: Least-squares frequency analysis of unequally spaced data. Astrophysics and Space Science 39, 447–462 (1976)
23. Norman, G., Streiner, D.L.: Biostatistics: The bare essentials, 3rd edn. People's Medical Publishing house, Shelton (2008)
24. Smith, W.D., Dutton, R.C., Smith, N.T.: Measuring the performance of anesthetic depth indicators. Anesthesiology 84, 38–51 (1996)
25. Kim, J.O.: Predictive Measures of Ordinal Association. The American Journal of Sociology 76, 891–907 (1976)

Time Series Prediction for Energy-Efficient Wireless Sensors: Applications to Environmental Monitoring and Video Games

Yann-Aël Le Borgne and Gianluca Bontempi

Machine Learning Group, Computer Science Department, CP212,
Faculty of Sciences, Université Libre de Bruxelles,
Bd Triomphe, Brussels, 1050, Belgium
{yleborgn,gbonte}@ulb.ac.be
http://mlg.ulb.ac.be

Abstract. Time series prediction techniques have been shown to significantly reduce the radio use and energy consumption of wireless sensor nodes performing periodic data collection tasks. In this paper, we propose an implementation of exponential smoothing, a standard time series prediction technique, for wireless sensors. We rely on a framework called *Adaptive Model Selection* (AMS), specifically designed for running time series prediction techniques on resource-constrained wireless sensors. We showcase our implementation with two demos, related to environmental monitoring and video games. The demos are implemented with TinyOS, a reference operating system for low-power embedded systems, and TMote Sky and TMote Invent wireless sensors.

Keywords: Wireless sensors, energy-efficiency, machine learning, time series prediction, exponential smoothing.

1 Introduction

Wireless sensor measurements typically follow temporal patterns, which are well approximated by time series prediction techniques. Different approaches have been proposed in the literature to approximate, by means of *parametric predictive models*, the measurements collected by wireless sensors [5,9,13,15]. The rationale of these approaches is that, if the parametric predictive model follows the sensor's measurements with sufficient accuracy, then it is enough to communicate the parameters of the model instead of the real measurements. In [7], a generic framework called *Adaptive Model Selection* was proposed, which encompassed previously proposed approaches based on time series prediction for wireless sensors. AMS was shown to provide, for a wide range applications, significant communication and energy savings.

In this paper, we investigate the use of exponential smoothing techniques in the AMS framework. Exponential smoothing (ES) is a standard time series prediction technique, known to perform well in many real-world applications [12]. We implement ES in TinyOS [14], a reference operating system for low power

F. Martins, L. Lopes, and H. Paulino (Eds.): S-Cube 2012, LNICST 102, pp. 63–72, 2012.
© Institute for Computer Sciences, Social Informatics and Telecommunications Engineering 2012

embedded systems, on TMote Sky and TMote Invent sensors [10]. We show that ES can be efficiently implemented, using negligible memory and computational requirements. We showcase our implementation with two different demos. The first demo implements a simple environmental monitoring system. A Java interface displays approximated light measurements collected by a wireless node. The second is a labyrinth game, where the player controls a ball on a 3D board by means of wireless inclinometers. We show that up to 90% of communication savings can be achieved using ES and AMS.

We summarize the rationale of the AMS framework in Section 2, and present how exponential smoothing can be implemented in the AMS in Section 3. Section 4 details the environmental monitoring and video gaming implementations.

2   Adaptive Model Selection

Adaptive Model Selection (AMS) [6,7] aims at reducing the radio and energy use of wireless sensors performing periodic data collection tasks, by using predictive models which approximate the real measurements.

Periodic data collection is typical of many wireless sensors' applications. For example, in environmental monitoring applications, the user's interest lies in the evolution of some physical quantity (temperature, humidity, light, ...) at periodic, fixed intervals. In gaming applications such as Wii games, the accelerometer embedded in the Wii controller device sends its measurements to the console at a predefined and high sampling rate. Denoting by $x[t]$ the measurements collected by a sensor at time t, we illustrate periodic data collection in Figure 1. A wireless sensor collects measurements at a predefined sampling rate, and sends them to a *base station*, i.e., a high-end computing unit which will get the same measurements as the sensor. Depending on the applications, the measurements are either displayed to the user (environmental monitoring), or processed for further actions (video gaming).

In many applications, it is often enough to collect an approximation of the sensor measurements. For instance, in plant growth studies, ecologists report that it is sufficient to have an accuracy of $\pm 0.5°C$ and 2% for temperature and humidity measurements, respectively [1]. It is therefore not necessary for a sensor to transmit all its measurements. This is the rationale of AMS [7], where *models* are used to approximate the measurements collected by a wireless sensor by means of time series prediction techniques

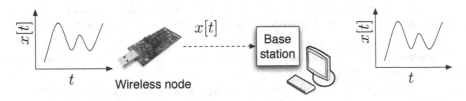

Fig. 1. Periodic data collection: all measurements $x[t]$ are transmitted to the base station, which gets the same measurements as those collected by the wireless sensor

In AMS, a *model* refers to a *parametric function* which predicts, at time $t+m$, $m \in \mathbb{N}^+$, the measurement of the sensor. Formally, the model is a function

$$h_\theta : \mathcal{X} \to \mathbb{R}$$
$$x \mapsto \hat{x}[t + m] = h_\theta(x)$$

where $x \in \mathcal{X}$ is the input to the model (typically a vector of measurements), θ is a vector containing the parameters of the model, and $\hat{x}[t+m]$ is the approximation of the model h to the measurement $x[t + m]$ at time $t + m$.

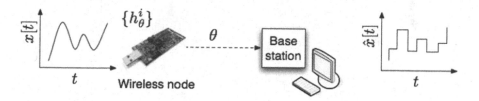

Fig. 2. Approximated data collection: the parameters θ of a predictive model are sent instead of the sensor measurements. Approximations of the measurements are obtained at the base station by means of the predictive model.

In AMS, the parameters of the model are sent instead of the measurements, as illustrated in Fig. 2. The models are estimated by the sensor node on the basis of its past measurements. Given an application-dependent error tolerance ϵ (for example $\epsilon = \pm 0.5°C$), the sensor node can locally assess if the prediction $\hat{x}[t+m]$ made by the model at time $t + m$ is within $\pm\epsilon$ of the true measurement $x[t+m]$. When the prediction is more than ϵ from the real measurement, a new set of model parameters is sent to the base station. At the base station, the model parameters are used to get approximations to the sensor measurements. AMS therefore **guarantees that all the approximated measurements obtained at the base station by means of predictive models are within $\pm\epsilon$ of the real measurements**.

As an example, let us detail the use of AMS with the constant model. The constant model is the most simple predictive model, and was proposed in [8,9]. The model assumes that the measurement at time $t + m$ is the same than that collected at time t, i.e.,

$$h_\theta : \mathcal{X} \to \mathbb{R}$$
$$x \mapsto \hat{x}[t + m] = x[t].$$

With a constant model, there is no model parameter, i.e., θ is empty. Fig. 3 illustrates how a constant model represents a temperature time series. The time series was obtained from the Solbosch Greenhouse of the University of Brussels, during a sunny summer day. Data were taken every 5 minutes, for a one day period, giving a set of 288 measurements. The measurements are reported with the red dashed lines, and the approximations obtained by a constant model with

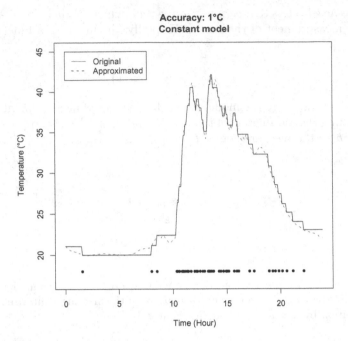

Fig. 3. A constant model acting for a one-day period on a temperature time series from the greenhouse at the University of Brussels, with a constant model and an error threshold set to $\epsilon = 1°C$.

an error threshold of $\epsilon = 1°C$ are reported with the black solid line. Updates are marked with black dots at the bottom of the figure. Using AMS, the constant model allows to reduce to 43 the number of measurements transmitted, resulting in about 85% of communication savings.

When the dynamics of the time series is not known *a priori* or in case of non-stationary signals, a set of models with different modeling abilities can be computed and assessed by a sensor node. The set of models is denoted by $\{h_\theta^i\}$, $1 \leq i \leq K$, where K is the number of models computed by the sensor node. The models are all assessed in parallel by the sensor node. When a model update is necessary, the parameters θ of the model that best approximates the sensor's measurements are sent to the base station.

3   Implementation of Exponential Smoothing in AMS

This section presents exponential smoothing (ES) and details its implementation in TinyOS and the adaptive model selection framework.

Exponential Smoothing

Exponential smoothing is a time series prediction technique [3] which has been shown to perform well for a wide variety of time series [2]. Different flavours of the technique have been proposed. The most simple one is the *simple exponential smoothing*, which consists in a weighted average of the past measurements. The weighted average is computed with

$$s[t] = \alpha x[t-1] + (1-\alpha)s[t-1]$$

where $0 \le \alpha \le 1$ is referred to as the *data smoothing factor*. Predictions are given with $\hat{x}[t+m] = s[t]$. Better approximations can usually be obtained using *double exponential smoothing*, with

$$\begin{aligned}
s[t] &= \alpha x[t-1] + (1-\alpha)(s[t-1] + b[t-1]) \\
b[t] &= \beta(s[t] - s[t-1]) + (1-\beta)b[t-1]
\end{aligned} \tag{1}$$

where $0 \le \beta \le 1$ is referred to as the *trend smoothing factor*. Predictions are obtained with

$$\hat{x}[t+m] = s[t] + mb[t]. \tag{2}$$

Note that the simple exponential smoothing is a particular case of double exponential smoothing, with $\beta = 0$. Exponential smoothing is computationally thrifty, which makes it suitable for implementation on resource-constrained wireless sensors.

Implementation in AMS

The use of exponential smoothing requires the specification of the values of the data and trend smoothing factors α and β. For our implementation, we chose to compute models with $\alpha \in \{0.2, 0.4, 0.6, 0.8, 1\}$ and $\beta \in \{0, 0.2, 0.4, 0.6, 0.8, 1\}$, so that the wireless node assesses a set of $K = 5 * 6 = 30$ models.

The parameters of the model is the couple $\theta = (s[t], b[t])$ (Eq. 2). Note that θ is sufficient to compute $\hat{x}[t+m]$, $m \in \mathbb{N}^+$, at the base station without knowledge of the true measurements $x[t]$ (See Eq. 2). The initialization is performed at time $t = 1$ by setting $s[1] = x[1]$ and $b[1] = 0$ both on the sensor node and at the base station, for each of the 30 models.

Models are assessed on the sensor node by means of the *relative update rate* [7], which is the average frequency at which packets are sent to the base station. Formally, let $U_i[t]$ be the update rate of model h_θ^i at time t, with $U_i[1] = 1$. When running AMS, the update rate of each model h_θ^i is updated at every time t with

$$U_i[t] = \frac{(t-1) * U_i[t-1] + 1}{t} \tag{3}$$

if model h_θ^i requires to send an update of its parameters θ, and

$$U_i[t] = \frac{(t-1) * U_i[t-1]}{t} \tag{4}$$

otherwise. The relative update rate reflects the percentage of transmitted packets with respect to periodic data collection. Note that the relative update rate for periodic data collection is 1 since it requires the transmission of the totality of

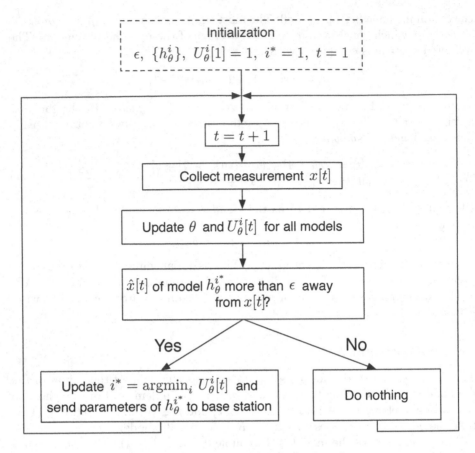

Fig. 4. Adaptive model selection algorithm

the measurements, and that any lower value indicates a gain in the number of transmitted packets. At time t, the best performing model denoted by $h_\theta^{i^*}$ is the one which minimizes the relative update rate, i.e., $i^* = \mathrm{argmin}_i\, U_i[t]$.

At runtime, only one model is shared between the sensor node and the base station. In our implementation, the first shared model is arbitrarily set to $i^* = 1$, i.e., the first model of the collection. Once the approximation for this model is $\pm\epsilon$ away from the real measurement, the node updates i^* to the model which minimizes the relative update rate, and sends the parameters of that model to the base station. A summary of the algorithm is given in Fig. 4.

Energy Efficiency

The rationale of AMS is to compute predictive models in order to reduce radio communication. We motivate in the following that the energy cost incurred by the computation of the predictive models is negligible compared to the energy saved by reducing communication.

The ratio of the energy spent in sending one bit of information to the energy spent in executing one instruction has been estimated to be around 2000 for a variety of real sensor network platforms [11].

Let us first assess the computational overhead of AMS. In the proposed implementation, the sensor node runs and estimates the relative update rate of 30 exponential smoothing models. This requires the sensor node to update, at each time t, the models' parameters (Eq. 1), as well as their relative update rates (Eq. 3 or 4). For one exponential smoothing model, the computational costs of these updates is very low, i.e., on the order of 20 multiplications and additions, depending on the implementation details. For 30 models, the computational overhead of AMS is therefore on the order of 600 CPU cycles.

Let us now assess the communication cost of a packet transmission. We relied in our implementation on TinyOS, a reference operating system for wireless sensor nodes. The standard TinyOS packet structure is detailed in Table 1. The packet overhead, i.e., the extra bytes of information, is 16 bytes. With AMS and exponential smoothing, the packet contains the model parameters $s[t]$ and $b[t]$, each stored on two bytes, giving a total packet size of 20 bytes.

Table 1. Detail of the TinyOS packet structure [14], with the size of each packet component (in bits). The packet overhead is 128 bits, i.e., 16 bytes.

Length	fcfhi	fcflo	dsn	dest	addr	type	group	Packet content	str	lqi	crc	ack	time
8	8	8	8	16	16	8	8	Model size	8	8	8	8	16

Sending a packet of 20 bytes (160 bits) therefore amounts to $160 * 2000 = 320000$ CPU cycles, i.e. running AMS over about $320000/600 \approx 530$ time instants. This means that if one of the shared predictive model is correct only once every 530 time instants, then AMS provides energy gains. This rough estimate suggests that the energy cost related to running the AMS is in most cases largely compensated by the energy gained in communication savings.

4   Demos and Implementation Details

This section presents two demos in which we apply our implementation of AMS with exponential smoothing. The first demo consists in an environmental monitoring application, and the second a video gaming application. The code for the demos is made available at [16].

Demo 1: Environmental Monitoring

In this demo, a TMote Sky wireless sensor, based on the TelosB prototype platform for research [10], is used to collect light measurements with ES and AMS. The sensor node is programmed in TinyOS v2.1 [14]. A demonstration mode is implemented, allowing to retrieve both the real measurements and the model parameters on a laptop or desktop computer. The measurements and approximations are displayed by means of a Java interface, see Fig. 5.

The snapshot shows the variations of light measurements (in Lux) in an office exposed to sunlight, for a one minute period, during which the sensor followed a 360 degree rotation around a vertical axis. The sampling rate was of 8Hz (i.e., around 60*8=480 measurement were collected), and the error tolerance $\max_t |\hat{x}[t] - x[t]| = 10$ Lux. AMS identified that, out of the 30 models, $\alpha = 0.2$ and $\beta = 0.6$ provided the most adapted smoothing factors. During that one minute period, almost 90% of the communications could be saved (i.e., around 50 model updates), while providing a qualitatively good approximation of the real measurements as can be seen on the user interface.

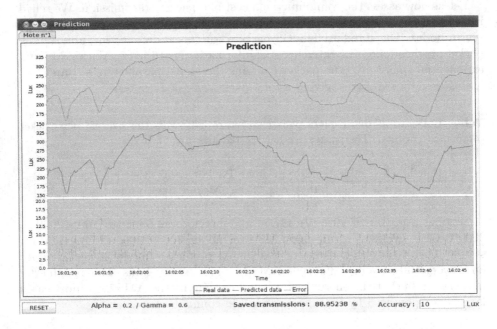

Fig. 5. Environmental monitoring interface displays light measurements collected by a TMote Sky sensor. Top chart shows real measurements, middle chart shows approximated measurements ($\epsilon = 10$ lx), bottom chart shows the predictive model error.

The stripped down version of the TinyOS code for periodic data collection takes 16546 bytes of RAM and 531 bytes of ROM. With AMS, the size of the code increases to 20334 bytes of RAM and 1155 bytes of ROM. AMS was therefore implemented using very little overhead in terms of computational resources.

Demo 2: Video Gaming

In this demo, a TMote Invent sensor node, also based on the TelosB design [10], is used to control a video game with a dual axis wireless inclinometer. The video game is a 3D labyrinth standing on a virtual board (see Fig. 6), where the goal

Fig. 6. 3D labyrinth game, controlled by TMote Invent inclinometers

is for the player to move the ball through the labyrinth while avoiding the holes. The game is implemented with Java 3D, and available as open source at [4]. The dual axis inclinometer measurements are used to control the orientation of the board. The node is programmed using Boomerang OS, a TinyOS version specifically designed for TMote nodes [14].

The error tolerance was fixed at 1°, so that approximations did not impair the fluidity and playability of the game. The amount of communications which can be saved depend on how fast the player moves the TMote Invent. Interestingly, since moving the ball through the labyrinth requires to gently control the sensor, we observed that significant communication savings could be saved, reaching more than 90% in most of the games played.

5   Conclusion

In this paper, we discussed an implementation of exponential smoothing in the adaptive model selection framework. The approach allows to perform approximated periodic data collection by relying on time series prediction models. We showed that the method can be implemented using little computational resources, and that significant communication and energy savings could be obtained in practical settings. We illustrated its use on two demos showcasing environmental monitoring and video gaming. In both cases, around 90% communication savings were obtained.

Acknowledgements. This work was supported by the ICT4Rehab project, sponsored by Brussels Institute for Research and Innovation, Innoviris, Belgium. Authors would like to thank the BA3 students of the University of Brussels for their contributions in designing the demos.

References

1. Deshpande, A., Guestrin, C., Madden, S., Hellerstein, J., Hong, W.: Model-driven data acquisition in sensor networks. In: VLDB 2004, pp. 588–599 (2004)
2. Gardner Jr., E.S.: Exponential smoothing: The state of the art. Journal of Forecasting 4(1), 1–28 (1985)
3. Holt, C.C.: Forecasting seasonals and trends by exponentially weighted moving averages. International Journal of Forecasting 20(1), 5–10 (2004)
4. Labyrinth 3D game. Java source code. Project Website,
   http://www.javafr.com/codes/LABYRINTHE-BILLE-JAVA3D_32818.aspx
5. Lazaridis, I., Mehrotra, S.: Capturing sensor-generated time series with quality guarantee. In: ICDE 2003, pp. 429–440 (2003)
6. Le Borgne, Y.: Learning in Wireless Sensor Networks for Energy-Efficient Environmental Monitoring. PhD thesis, ULB, Brussels, Belgium (2009)
7. Le Borgne, Y., Santini, S., Bontempi, G.: Adaptive model selection for time series prediction in wireless sensor networks. Journal of Signal Processing 87(12), 3010–3020 (2007)
8. Olston, C., Jiang, J., Widom, J.: Adaptive Filters for Continuous Queries over Distributed Data Streams. In: SIGMOD 2003, pp. 563–574 (2003)
9. Olston, C., Loo, B.T., Widom, J.: Adaptive precision setting for cached approximate values. ACM SIGMOD Record 30, 355–366 (2001)
10. Polastre, J., Szewczyk, R., Culler, D.: Telos: enabling ultra-low power wireless research. In: IPSN 2005, pp. 364–369 (2005)
11. Raghunathan, V.S., Srivastava, C.S.P.: Energy-Aware Wireless Microsensor Networks. IEEE Signal Processing Magazine 19(2), 40–50 (2002)
12. Santini, S.: Adaptive sensor selection algorithms for wireless sensor networks. PhD thesis, ETH Zurich, Zurich, Switzerland (2009)
13. Santini, S., Römer, K.: An adaptive strategy for quality-based data reduction in wireless sensor networks. In: INSS 2006, Chicago, IL, USA, pp. 29–36 (2006)
14. TinyOS. An Open-Source Operating System Designed for Wireless Embedded Sensor Networks. Project Website, http://www.tinyos.net
15. Tulone, D., Madden, S.: PAQ: Time Series Forecasting for Approximate Query Answering in Sensor Networks. In: Römer, K., Karl, H., Mattern, F. (eds.) EWSN 2006. LNCS, vol. 3868, pp. 21–37. Springer, Heidelberg (2006)
16. Wireless Lab - University of Brussels. Code and datasets. Project Website,
   http://www.ulb.ac.be/di/labo/datasets.html

Connected Cars: How Distributed Data Mining Is Changing the Next Generation of Vehicle Telematics Products

Hillol Kargupta[1,2]

[1] Agnik
[2] Computer Science & Electrical Engineering Department,
University of Maryland, Baltimore County
hillol@agnik.com

Abstract. Modern vehicles are embedded with varieties of sensors monitoring different functional components of the car and the driver behavior. With vehicles getting connected over wide-area wireless networks, many of these vehicle diagnostic-data along with location and accelerometer information are now accessible to a wider audience through wireless aftermarket devices. This data offer rich source of information about the vehicle and driver performance. Once this is combined with other contextual data about the car, environment, location, and the driver, it can offer exciting possibilities. Distributed data mining technology powered by onboard analysis of data is changing the face of such vehicle telematics applications for the consumer market, insurance industry, car repair chains and car OEMs. This talk will offer an overview of the market, emerging product-types, and identify some of the core technical challenges. It will describe how advanced data analysis has helped creating new innovative products and made them commercially successful. The talk will offer a perspective on the algorithmic issues and describe their practical significances. It will end with remarks on future directions of the field of Machine-to-Machine (M2M) sensor networks and how the next generation of researchers can play an important role in shaping that.

Brief Biography. Dr. Hillol Kargupta is a Professor of Computer Science at the University of Maryland, Baltimore County. He is also a co-founder of AGNIK, a vehicle performance data analytics company for mobile, distributed, and embedded environments. He received his Ph.D. in Computer Science from University of Illinois at Urbana-Champaign in 1996. His research interests include mobile and distributed data mining. Dr. Kargupta is an IEEE Fellow. He won the IBM Innovation Award in 2008 and a National Science Foundation CAREER award in 2001 for his research on ubiquitous and distributed data mining. He and his team received the 2010 Frost and Sullivan Enabling Technology of the Year Award for the MineFleet vehicle performance data mining product and the IEEE Top-10 Data Mining Case Studies Award. His other awards include the best paper award for the 2003 IEEE International Conference on Data Mining for a paper on privacy-preserving data mining, the 2000 TRW Foundation Award, and the 1997 Los Alamos Award for Outstanding Technical

F. Martins, L. Lopes, and H. Paulino (Eds.): S-Cube 2012, LNICST 102, pp. 73–74, 2012.
© Institute for Computer Sciences, Social Informatics and Telecommunications Engineering 2012

Achievement. His dissertation earned him the 1996 Society for Industrial and Applied Mathematics annual best student paper prize. He has published more than one hundred peer-reviewed articles. His research has been funded by the US National Science Foundation, US Air Force, Department of Homeland Security, NASA and various other organizations. He has co-edited several books. He serve(s/d) as an associate editor of the IEEE Transactions on Knowledge and Data Engineering, IEEE Transactions on Systems, Man, and Cybernetics, Part B and Statistical Analysis and Data Mining Journal. He is/was the Program Co-Chair of 2009 IEEE International Data Mining Conference, General Chair of 2007 NSF Next Generation Data Mining Symposium, Program Co-Chair of 2005 SIAM Data Mining Conference and Associate General Chair of the 2003 ACM SIGKDD Conference, among others.

Modelling and Simulation
of Underwater Low-Power Wake-Up Systems

Salvador Climent, Antonio Sanchez, Juan Vicente Capella,
Sara Blanc, and Juan José Serrano

Universitat Politècnica de València, Institut ITACA,
Camí de Vera s/n, 46022 Valècia, Spain
{scliment,ansanma7,jserrano}@itaca.upv.es,
{jcapella,sablacla}@disca.upv.es

Abstract. Underwater Wireless Sensor Networks (UWSN) have become
an important area of research due to its many possible applications. One
example are the long-term monitoring applications were the nodes only
need to be awake during a small fraction of time. This kind of applications
can greatly benefit from low-power, wake-up systems. However, despite
the fact that the simulations can greatly improve the development time
of new algorithms and features, optimizing their performance, up until
today there is no wake-up system model available.

In this paper a low-power underwater wake-up model for the ns-3
simulator is going to be presented. Using this model the, as far as we
know, only two available underwater modems with integrated wake-up
capabilities are compared in terms of energy consumption.

Keywords: Underwater   modems,   energy-efficiency,   wake-up,
modelling, simulation.

1   Introduction

Underwater Wireless Sensor Networks (UWSN) have become an important area
of research due to its wide range of applications, ranging from submarine surveillance to monitoring of the marine environment. Recent advances [3] have made
feasible to develop relatively large underwater networks although the deployment
and maintenance costs are still high. Hence, there is still a need for low-cost, low-
power modems capable of extending the nodes' battery life as much as possible.

This is usually the case of long-term monitoring applications, were the nodes
only need to be awake during a small fraction of time and remain in the sleep
state for most of the time. To that end, underwater low duty-cycle MAC protocols have been develop in order to allow the nodes to remain in the sleep state
as much as possible and implement some degree of synchronization to maintain
the connectivity of the network.

These approaches usually maintain a permanently and rotative awake
backbone to assure the connectivity of the network at all times or periodically wake-up the nodes so they can hear for transmissions intended for them.

F. Martins, L. Lopes, and H. Paulino (Eds.): S-Cube 2012, LNICST 102, pp. 75–88, 2012.

This solutions, although very interesting, maintain some nodes awake during periods of time where no data transmission is being made.

Nowadays acoustic modem design allows the implementation of a low-power reception state where nodes keep listening the channel with very low power consumption (much lower than the regular reception state), while the not necessary circuitry remains in sleep state [17], [10]. This way, the modem is able to listen and recognize certain stimuli sent prior to the actual data packet and wake-up the main circuitry to receive it. In [4] the authors present a study where the viability in terms of energy-efficiency of this solutions is shown.

Since the deployment costs of this networks is very high, simulations are an essential tool to test and tune new features and algorithms before their implementation with real hardware. However, to the best of our knowledge, there is no underwater wake-up model available.

In this paper a low-power underwater wake-up model for the ns-3 simulator [7] is going to be introduced. Using this model the, as far as we know, only two available underwater modems with integrated wake-up capabilities are going to be compared in terms of energy consumption.

The remaining of this paper is organized as follows, in Section 2 the more relevant MAC protocols based on the previously introduced sleep-cycle solutions are introduced. In Section 3 the main features and energy consumption characteristics of the available underwater wake-up modems will be presented. Section 4 introduces the proposed underwater wake-up model for the ns-3 simulator. Section 5 describes our simulations and obtained results. Finally, in Section 6 conclusions are drawn and future work in this line is highlighted.

2   Related Work

The design of energy efficient and reliable communication protocols has been one of the main research focus in radio-frequency (RF) sensor networks and in underwater wireless sensor networks. The key aspect in order to accomplish this, is to design a medium access protocol (MAC) that is able to achieve good packet delivery ratios while consuming as less energy as possible.

In order to prolong the battery life, different methods can be employed from the medium access perspective. One of them is to reduce the packet signalling overhead and lower the number of collisions. Another one is to switch off the radio transceiver when the node does not need to send or receive any data. However, by doing so, the node is totally disconnected from the network and can no longer act as a relay node if needed.

There are different approaches on how to disconnect this nodes and bring them back online. The Geographic Adaptive Fidelity (GAF) [19] achieves this by dividing the sensor network into small grids in such a way that at each point in time only one node in the grid can be in active mode, while the others have their radios disconnected. This protocol, originally developed for RF networks might be difficult to apply underwater, since it needs some location information and partial synchronization between the nodes. Moreover, it relays on dense networks, which might be difficult to achieve underwater, due to the high node and deployment costs.

Span [2] is another protocol developed originally for RF networks, which selects a limited set of nodes to form a backbone where the packets can be forwarded at all times. To form this backbone, the protocol needs to interact with the routing layer. The other nodes can transition to sleep states more frequently since the do not have to relay any data. Moreover, in order to distribute the energy consumption, the nodes are rotated in their backbone role. The performance of this algorithm is very density-dependant so it might be difficult to achieve high energy savings in an underwater environment.

The protocol called STEM (Sparse Topology and Energy Management) [14] takes a different approach than the other two previously introduced. While GAF and Span try to always maintain a path for the packets to be forwarded without inquiring in long packet delays, STEM trades off this latency for energy savings. To do so, all nodes are in a low duty-cycle sleep state and when a sender wants to transmit, it first send a beacon in such a way that it is guaranteed to be received by the receiver in some bounded time. When the receiver wakes up and hears the beacon it informs the sender and gets prepared to receive the data. Although the authors claim that it is not necessary, the results in [14] are obtained under the assumption that there are two radio transceivers available, with the associated extra cost and consumption, one for the actual data transmission and one for the transmission and reception of the beacon signals.

Although they can be applied underwater, this protocols where specifically design for RF networks. Given the harsh underwater environment and the different transmission characteristics [3], there has been some effort from the scientific community in order to design new protocols for the underwater environment.

Since in the underwater acoustic environment the cost of transmitting one data packet is much higher than the cost of receiving it, in [20] the authors analyze the effect, form the energy-efficiency point of view, of using low duty-cycle protocols in the underwater networks. They show that a simple duty-cycle scheme, coupled with some power control can reduce the overall energy spent in the network.

R-MAC [18] is one of this low duty-cycle protocols specifically designed for underwater networks. In this protocol each node periodically transits between the listen and sleep states. The duration of this states is the same for all nodes and each one randomly selects when to perform this transition. After each node estimates the latency to all its neighbours it starts broadcasting its own schedule and learning the others, until all the nodes in the network have the information from all their neighbours. After this configuration phase the nodes can start to transition between their wake-up and sleep states and to send their data. In order to send data packets, the sender have to agree with the receiver on when to do this transmission and it has to be done during a reserved period of time called R-windows. The protocol does need to re-run its configuration phase so nodes can synchronise again and this can be very energy consuming depending on the actual time they need to perform the configuration.

Another low duty-cycle protocol specifically design for underwater networks is the UWAN-MAC [8]. This protocol sets up some sort of adaptive TDMA (Time Division Multiple Access [15]) where each node broadcasts when is it going to send its data and learns the schedule of the others in order to wake-up for the transmissions on which it is acting as the receiver. After that, on each data transmission the node piggybacks when is going to be the next transmission. This way, the receiver can adapt to the new schedule. This protocol assumes that the clock drift is not significant and that the sound speed will remain constant between the schedule updates.

In this kind of protocols, where nodes have to partially synchronise themselves in order to be able to achieve a good operation, the studies some times fail to quantify the cost of the network reconfiguration and the clock drifts might have in the overall performance and energy consumption. As an example, Casari et al. analyze in [1] how fast the network performance drops for increasing synchronization drifts when using the UWAN-MAC protocol.

Another approach for the nodes to efficiently transition between the wake-up and sleep states comes with the introduction of a low-power, wake-up modem. This wake-up system consumes very low energy while it remains on the reception state so, a node can turn off its main radio and remain listening on this low-power radio waiting to receive a transmission.

In [4] the authors analyze the situations in which a low-power, wake-up modem can save energy compared to idle-time sleep cycling algorithms. They show that, in fact, wake-up modems can outperform the sleep cycling solutions and even behave almost as well as the ideal sleep cycle where nodes magically wake-up exactly at the same time when they have to receive a packet.

In the following sections, the currently available low-power, wake-up modems are going to be introduced and compared using the proposed underwater low-power, wake-up system model for the ns-3 simulator.

3   Low-Power, Wake-Up Modems

To the best of our knowledge, currently there are two proposed modems with an integrated wake-up system: the Wills underwater modem [17] and the ITACA modem [11], [10], [12].

3.1   Wills Underwater Modem

Wills underwater modem is a low-power acoustic modem designed for small range networks [17]. It includes a dedicated wake-up tone receiver and a Mica mote for the packet processing and coding, this motes consume when activated $24mW$[5]. Its communication distance ranges from $50m$ up to $500m$ using a frequency-shift keying (FSK) modulation at $1kbps$. The wake-up signal is sent in the same frequency band as the regular signal. This modem consumes at most $2W$ when transmitting, $44mW$ in reception state and $500\mu W$ in sleep state, including the wake-up circuitry consumption.

3.2   ITACA Modem

The ITACA modem has been designed as an energy-efficient architecture for small/ medium range networks, with low-power UWSN consumption restraints [10]. Its architecture is based on a microcontroler (MCU) that only consumes $24mW$ in reception and $3\mu W$ in sleep state. It is capable of transmitting upto $100m$ using a frequency-shift keying (FSK) modulation at $1kbps$ and consuming $120mW$.

The acoustic wake-up signal is transmitted using an on-off keying (OOK), which is compatible without additional hardware, with the FSK modulation used for the regular transmissions. Hence, this modem also transmits the wake-up signal in the same frequency band as the regular signal.

In order to be able to handle acoustic wake-up signals, the ITACA modem includes an off-the-shelf commercial peripheral, the AS3933 from Austria Microsystems [6]. Since this circuit is intended to be triggered using magnetic coupled signals, a net was specifically designed to adapt the acoustic incoming signals to the RFID based wake-up circuit. This IC with the adaptation net consumes $8.1\mu W$.

Aside from the wake-up tone capability, this modem also includes de possibility, with the same energy consumption, to program and send different wake-up patterns allowing to perform a selective wake-up and only activating the receiving node.

4   The ns-3 Simulator and the Wake-Up Model

The ns-3 is a discrete event simulator for computer and sensor networks. The simulator is organized in modules as shown in Figure 1. The core module offers different libraries to help on the development of models and protocols like smart pointers, trace system, callback objects, etc. [7].

The common module aggregates the data types related to network packets and their headers and the module simulator deals with the simulation time and the different schedulers. Going up though the stack, the node module defines a node in the ns-3 simulator and other different classes. The module mobility is in charge of modelling the mobility of the nodes (in this work the nodes are going to be static).

Following up, the routing, internet-stack and devices modules are the ones that implement the routing protocols like AODV or OSLR; the TCP/IP stack and the devices like Ethernet cards.

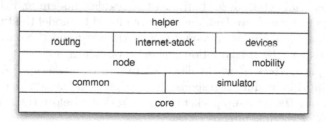

Fig. 1. ns-3 modules

Finally the helper module includes different class definitions implementing an API to facilitate the script programming and the configuration of the simulation to the user.

4.1 Packet Structure

Two different packets are used by the proposed model. The UanPacket, which is included in the standard ns-3 distribution, contains 1 byte for the sender address, 1 byte for the receiver address and 1 byte for the type of the payload. And a newly implemented packet named UanPacketWU, which contains 1 byte for the tone and wake-up pattern.

4.2 Wake-Up Modem Model

In this section the model implementation of this low-power, wake-up system is discussed. As stated before, this model is inspired on the low-cost, low-power, wake-up system by A. Sanchez et al. [11], [10], [12] and the Wills modem [17]. These two modems have similar block diagrams hence, it is easy to compare them in terms of energy consumption by adjusting the consumption power of the different parts.

A simplified component diagram of the ns-3 underwater modem is depicted in Figure 2. The UanNetDevice component models a Network Interface Card (NIC) and is used by the upper layers to send and receive packets to/from the network. The UanMac component models the medium access protocol used by the nodes in the configured network. The ns-3 has some MAC protocols already implemented and ready to use like CW-MAC [9] or the widely known Simple ALOHA.

The UanPhy component models the underwater physical layer and it includes different Signal-to-Noise-Interference Ratio (SNIR) and Packet Error Rate (PER) models. Finally, the UanChannel module models an underwater channel and delivers all the packets to the UanPhy components connected or listening to this specific underwater channel.

The UanPhy component is also connected to a DeviceEnergyModel, which models the energy consumption of the different radio states and, in turn, is connected to an EnergySource, which models different energy containers like batteries.

The presented Wake-Up model is depicted in Figure 3. Like in the regular model, there is one UanNetDevice acting as a NIC card and one UanChannel to emulate the underwater channel. In order to model the regular modem and the wake-up system, two UanPhy and two UanMac are introduced to model the two systems.

The two UanPhy modules are equal in functionality and the only difference between the two is the consumption parameters, which are set to match the consumption parameters of the regular modem and the wake-up system.

The UanMacWU module is responsible for doing the actual channel assessment and sending the wake-up packet (UanPacketWU) before the actual packet that the UanMac module intends to send.

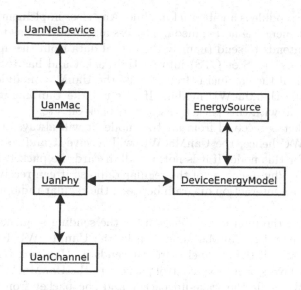

Fig. 2. Simplified component diagram of ns-3 underwater modem

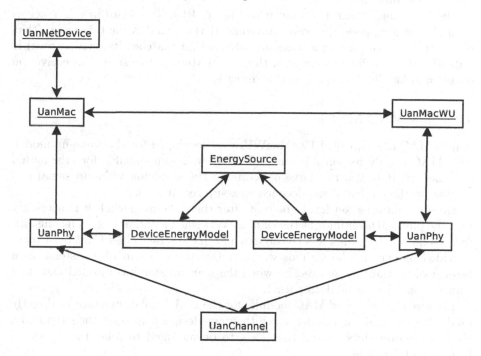

Fig. 3. Simplified component diagram of the proposed underwater wake-up model

The UanMac module is an abstract module. Any class implementing this module should implement the actual medium access algorithm. It has to encapsulate the data that intends to send (namely the actual data from the upper layers or control packets like RTS or CTS) into an UanPacket and has to ask the UanMacWU module if the channel is free. If it is, the UanMac module has to send the packet to the UanMacWU module. If it is not the UanMac module has to decide what to do with this packet, discard it, back-off, etc.

When a packet is received from another node, it will always be preceded by an UanPacketWU hence, the UanMacWU will receive it and see if the packet destination is for this node. If it is not, it will discard the packet. If it is, it will wake-up the UanPhy in charge of the regular radio so it can receive the packet. Figure 4 shows an interaction diagram between the regular radio model and the wake-up model.

In Figure 4(a) the interaction diagram for the sending sequence is depicted. It can be seen how the UanMac module asks the UanMacWU for the current channel status and, if the channel is free, it sends its UanPacket with the data form the upper layers or its own control packets to the UanMacWU. After that, the UanMacWU sends the wake-up packet and the packet from the UanMac immediately after it.

The receiving sequence is shown in Figure 4(b). The UanPhyWU receives from the sending node the wake-up signal. If the signal is for this node, (that is, it is a wake-up tone or a wake-up pattern that matches its own pattern) it wakes-up the UanPhy associated with the UanMac module so it can receive the packet and handle it to the UanMac directly.

4.3   New MAC Models

A new MAC model, called UanMacWU, was developed for the wake-up model. This MAC is only responsible for sending the wake-up signal before the actual data packet. It is also in charge of receiving the incoming wake-up signal and waking up the modem if the decoded wake-up requires so.

Since the data packet has to be sent after the wake-up packet, it is necessary that the physical layer informs the UanMacWU when a packet has been transmitted. To do so, the UanMacWU also extends the abstract class UanPhyListener provided by the simulator. This way, the UanMacWU can be registered as a listener of its UanPhy and receive when the transmission has finished and start transmitting the actual data packet.

The already developed MAC modules for the ns-3 simulator cannot be directly used with this wake-up model, since they are designed to send their data to a UanPhy module directly and they have to be modified to relay the data to a UanMacWU module.

Another UanMac module was developed in order to test our model. This new MAC module implements an ALOHA-CS (ALOHA with Carrier Sensing) with a FIFO queue for the outgoing packets and a configurable back-off timer.

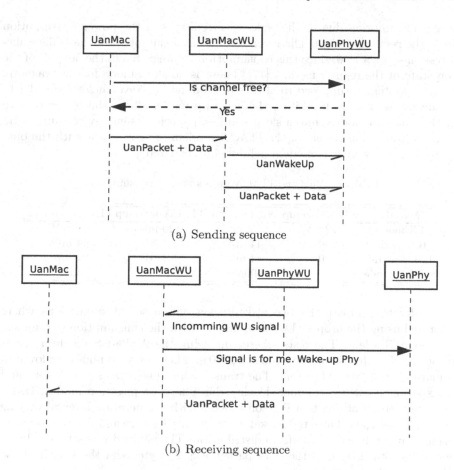

(a) Sending sequence

(b) Receiving sequence

Fig. 4. Interaction diagram between the radio model and the wake-up model

5   Wake-Up Systems Evaluation

In this section, an evaluation of the two wake-up systems introduced in Section 3 is performed using the proposed wake-up model for the ns-3 simulator.

To that end, the first step has to calculate the transmission state power consumption of the Wills modem. In the original paper [17] the authors only specify the energy consumption when transmitting full-power. Since the ITACA modem is able to transmit upto 100 m, in order to perform a fair comparative, we estimated that the energy consumption of the Wills modem when transmitting to 100 m is 172 mW. Details on how we estimated this value can be found in the appendix.

The other energy consumption parameters used from the Wills modem along with the ones from the ITACA modem can be found in Table 1 and where extracted from [17] for the Wills modem and [10] for the ITACA modem. Although this modems integrate the wake-up systems, this table differentiates between the energy consumption due to the wake-up circuitry and the power consumption due to the regular modem. The reception mode consumption of the Wills wake-up system was set to 0, since this consumption is integrated in the $500\mu W$ of the sleep state of the regular modem [17]. There is no sleep mode for the wake-up, since this system is supposed to always remain awake. Nevertheless, in order to provide the maximum flexibility, the wake-up model allows the researchers to put the wake-up module into a sleep state and specify its energy consumption.

The consumption results of the ITACA modem are consistent with the ones reported in [13] were this modem is fully characterized.

Table 1. Wills and ITACA radios energy consumption

MODE	Wills Wake-up	Wills modem	ITACA wake-up	ITACA modem
TX mode	172 mW	172 mW	120 mW	120 mW
RX mode	0	44 mW	$8.1\mu W$	24 mW
IDLE mode	0	44 mW	$8.1\mu W$	24 mW
SLEEP mode	-	$500\mu W$	-	$3\mu W$

In order to compare the two wake-up systems a set of simulations where performed using the proposed wake-up model with the consumption parameters specified in Table 1. The tests where run using the Aloha-CS medium access protocol and with a simulation time of 1000 s. There were 10 nodes deployed in a square area of 20×20 meters. The transmission speed was set to 1 kbps and the experiments where conducted using the Thorp propagation model. Traffic was generated according to a Poisson process with a generation interval varying between 1 and 100 s. Once the packet was generated source and destination were selected randomly using a uniform distribution. The payload was set to 20 bytes hence, the total length of the data packet was 23 bytes and the length of the wakeup packet 1 byte.

Figure 5 shows the results of the simulations. One can notice how the ITACA modem using the wake-up mode tone outperforms Wills modem by saving 54% more energy with the lowest traffic rate (Wills 1.83 J, ITACA 0.83 J) and a 41% with the highest traffic (Wills 55.08 J, ITACA 32.40 J). But, the highest performance gain comes with the wake-up pattern mode. Using this mode, only the intended receiver of the transmission wakes-up. This provides huge energy savings ranging from 74% upto 80%.

Comparing this pattern mode to the optimum wake-up (where there is no energy consumption from the wake-up circuitry or from transmitting the wake-up signal), Figure 5 shows how the ITACA modem with the pattern mode behaves. Results show how the wake-up energy consumption of the ITACA modem represents only an 8% more in the worst case.

One may want to compare these consumption values with the ones where there is no wake-up system. As a reference, the energy consumption of the ITACA modem when there is 100 seconds of traffic generation interval and there is no wake-up system available would be approximately 24 J, which is much higher than the 0.83 J achieved with the wakeup tone.

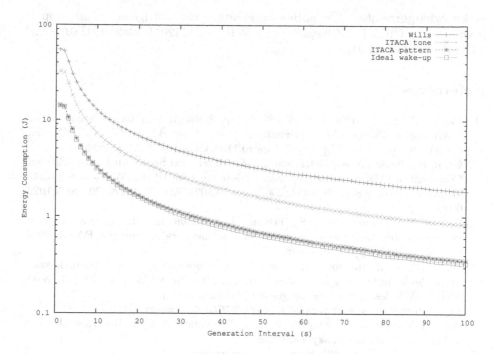

Fig. 5. Comparison results of the two different modems

6   Conclusions

Simulations have proven to greatly facilitate the research and development of new algorithms for underwater sensor networks. By using simulations tools, researchers can develop and test new algorithms before their real deployment, giving insights on the performance an algorithm might have in a real environment. Of course, the system to be simulated has to be as accurately modelled as possible to match the features of the real system.

In this paper, a model of an underwater wake-up modem has been presented. The model has been designed in order to match the most important features of the only two wake-up systems available today, but maintaining certain flexibility to be able to adapt it to new features the researchers might want to test.

Using this model, the two wake-up systems where tested and compared in terms of energy consumption to the optimum wake-up (the one in which the wake-up circuitry consumes nothing) and results showed that the ITACA modem can provide up to 80% energy savings when compared to the Wills modem and only spends 8% more energy than the optimum case.

Future work will include further development of the wake-up model adapting it to allow the transmission of the wake-up signal using different transmission channels and the research and development of new protocols that can take advantage of the wake-up functionality.

Acknowledgements. The authors gratefully acknowledge financial support from the CICYT (research projects CTM2011-29691-C02-01 and TIN2011-28435-C03-01).

References

1. Casari, P., Lapiccirella, F.E., Zorzi, M.: A Detailed Simulation Study of the UWAN-MAC Protocol for Underwater Acoustic Networks. In: Oceans 2007, Vancouver, BC, pp. 1–6. IEEE (2007), doi:10.1109/OCEANS.2007.4449226
2. Chen, B., Jamieson, K., Balakrishnan, H., Morris, R.: Span: An Energy-Efficient Coordination Algorithm for Topology Maintenance in Ad Hoc Wireless Networks. Wireless Networks 8(5), 481–494 (2002), doi:10.1023/A:1016542229220, ISSN 1022-0038
3. Chitre, M., Shahabudeen, S., Freitag, L., Stojanovic, M.: Recent advances in underwater acoustic communications & networking. In: OCEANS 2008, vol. 2008(suppl.), pp. 1–10 (2008)
4. Harris III, A.F., Stojanovic, M., Zorzi, M.: Idle-time energy savings through wake-up modes in underwater acoustic networks. Ad Hoc Networks 7(4), 770–777 (2009)
5. MICAz. Wireless Measurement System (January 2012), http://courses.ece.ubc.ca/494/files/MICAz_Datasheet.pdf.
6. Austria Microsystems. AS3933 3D Low Frequency Wakeup Receiver (January 2012), www.austriamicrosystems.com
7. Ns-3. ns-3 (January 2012), http://www.nsnam.org
8. Park, M.K., Rodoplu, V.: UWAN-MAC: An Energy-Efficient MAC Protocol for Underwater Acoustic Wireless Sensor Networks. IEEE Journal of Oceanic Engineering 32(3), 710–720 (2007), doi:10.1109/JOE.2007.899277, ISSN 0364-9059
9. Parrish, N., Tracy, L., Roy, S., Arabshahi, P., Fox, W.: System Design Considerations for Undersea Networks: Link and Multiple Access Protocols. IEEE Journal on Selected Areas in Communications 26(9), 1720–1730 (2008), doi:10.1109/JSAC.2008.081211, ISSN 0733-8716
10. Sanchez, A., Blanc, S., Yuste, P., Piqueras, I., Serrano, J.J.: A low-power wake-up system for underwater wireless sensor modems. In: Proceedings of the Sixth ACM International Workshop on Underwater Networks - WUWNet 2011, Seattle, Washington, USA, pp. 18:1–18:2. ACM Press (December 2011), doi10.1145/2076569.2076589, ISBN 9781450311519
11. Sanchez, A., Blanc, S., Yuste, P., Serrano, J.J.: RFID based Acoustic Wake-Up system for Underwater Sensor Networks. In: 2011 IEEE 8th International Conference on Mobile Adhoc and Sensor Systems (MASS), Valencia, Spain, pp. 873–878 (2011)
12. Sanchez, A., Blanc, S., Yuste, P., Serrano, J.J.: A low cost and high efficient acoustic modem for underwater sensor networks. In: OCEANS 2011 IEEE SANTANDER, Santander, Spain, pp. 1–10 (2011)
13. Sanchez, A., Blanc, S., Yuste, P., Serrano, J.J.: Advanced Acoustic Wake-up System for Underwater Sensor Networks. Communications in Information Science and Management Engineering 2(2), 1–10 (2012)
14. Schurgers, C., Tsiatsis, V., Ganeriwal, S., Srivastava, M.: Topology management for sensor networks. In: Proceedings of the 3rd ACM International Symposium on Mobile Ad Hoc Networking & Computing, MobiHoc 2002, Lausanne, Switzerland, pp. 135–145. ACM Press (June 2002), doi:10.1145/513800.513817, ISBN 1581135017

15. Tanenbaum, A.S.: Computer Networks. Prentice Hall (2003)
16. TI. Analog, Embedded Processing, Semiconductor Company, Texas Instruments (January 2012), www.ti.com
17. Wills, J., Ye, W., Heidemann, J.: Low-power acoustic modem for dense underwater sensor networks. In: WUWNet 2006 Proceedings of the 1st ACM International Workshop on Underwater Networks, pp. 79–85 (2006)
18. Xie, P., Cui, J.-H.: R-MAC: An Energy-Efficient MAC Protocol for Underwater Sensor Networks. In: International Conference on Wireless Algorithms, Systems and Applications (WASA 2007), pp. 187–198. IEEE (August 2007), doi:10.1109/WASA.2007.37, ISBN 0-7695-2981-X
19. Xu, Y., Heidemann, J., Estrin, D.: Geography-informed energy conservation for Ad Hoc routing. In: Proceedings of the 7th Annual International Conference on Mobile Computing and Networking, MobiCom 2001, Rome, Italy, pp. 70–84. ACM Press (July 2001), doi:10.1145/381677.381685, ISBN 1581134223
20. Zorzi, F., Stojanovic, M., Zorzi, M.: On the effects of node density and duty cycle on energy efficiency in underwater networks. In: OCEANS 2010 IEEE SYDNEY, Sydney, pp. 1–6. IEEE (May 2010), doi:10.1109/OCEANSSYD.2010.5603904, ISBN 978-1-4244-5221-7

Appendix

Wills modem transmitter is shown in Figure 6. From the data reported in [17], when the output power is set to 2 W the transmission can reach 500 m. Nevertheless, this is the maximum output level and the output power amplifier can be configured using four different gain values. The shorter communication distance, the lower power is needed to maintain the same communication performance. Therefore, to make a faithful comparison between the ITACA modem and Wills modem, the output power needs to be reduced to match 100 m.

To calculate the amount of power in transmission that the Wills modem needs to reach 100 m communications, the following statements are assumed:

1. Wills modem supports four discrete transmission power levels: 15dBm, 21dBm, 27dBm and 33dBm [17].
2. Output power should be near to 108 mW (20.3 dBm), which is the power reported in [12] to reach 100 m using ITACA modem.
3. TX power consumption is mainly due to two main blocks: power amplifier and node micro-controller. The rest of the modem elements are considered to be switched off or their power consumption is considered negligible.

$$P_{supply}(mW) = \frac{P_{out}(mW)}{\eta_{amplifier}} \tag{1}$$

The minimum value found to reach 100 m. was reported in [12] to be 20.3 dBm. Hence, we have to choose the higher value closer to this one in the Wills power amplifier, which is 21 dBm.

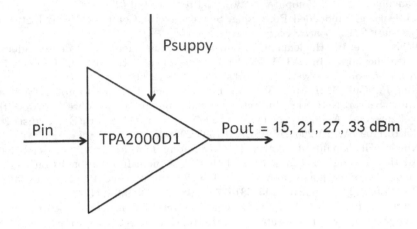

Fig. 6. Wills Modem acoustic wave transmitter diagram

The energy demanded to the power supply from the amplifier output power is obtained using expression (1). Wills output amplifier efficiency (η) is around 0.85 [16]. So, we estimate that Wills power consumption to transmit messages to nodes placed 100 m far is 148 mW (21.7 dBm). This value only accounts to the power amplifier, Wills modem architecture embeds a Mica, which power consumption in active mode corresponds to 24 mW [5]. Concluding, we estimate that Wills power consumption for 100 meter links in TX mode is 172 mW (22.3 dBm).

Collaborative Sensing Platform for Eco Routing and Environmental Monitoring

Markus Duchon[1], Kevin Wiesner[2],
Alexander Müller[2], and Claudia Linnhoff-Popien[2]

[1] Siemens Corporate Technology, CT T DE IT 1
Otto-Hahn-Ring 6, 80200 Munich, Germany
markus.duchon.ext@siemens.com
[2] Ludwig-Maximilian-University Munich, Mobile and Distributed Systems
Oettingenstr. 67, 82358 Munich, Germany
{kevin.wiesner,linhoff}@ifi.lmu.de, muelleral@cip.ifi.lmu.de

Abstract. During the past decades, ecological awareness has been steadily gaining popularity. Especially in so called Megacities, the burden caused by air pollution is very high, as millions of people live together in a localized manner. To be aware of the current pollution status, selective measuring stations where deployed in the past. The idea of this work is to enable the masses to participate in obtaining and using their own measurements, e.g. with future generations of mobile phones that are equipped with adequate sensors. The proposed platform allows for a high resolution environmental monitoring and provides additional services such as *Eco Routing* or visualization. Furthermore, we will present the results of the platform's performance as well as a comparison between the traditional (shortest/fastest) routing and the novel (shortest/fastest) *Eco Routing* approach.

Keywords: Collaborative sensing, monitoring platform, Eco Routing.

1 Introduction

During the past decades, environmental problems have been gaining more and more attention, especially in highly populated areas. In Europe, many administrations raised the bar to observe certain specified pollution limits within urban areas. For this purpose, environmental zones have been defined in most of the bigger cities in Germany. The entry to these zones is restricted to certain types of vehicles, e.g. those that are equipped with state of the art technologies to minimize exhaust emissions. But, according to a preliminary study[1] of the German Federal Environment Agency (UBA), the values of fine particulate matter in 2011 exceeded the average of the last four years, although these zones have been established. Besides traffic, also combustion processes in industry and private households attract increasing attention in the area of environmental observation.

---

[1] http://www.umweltdaten.de/publikationen/fpdf-1/4211.pdf,
last access 01.02.2012.

F. Martins, L. Lopes, and H. Paulino (Eds.): S-Cube 2012, LNICST 102, pp. 89–104, 2012.
© Institute for Computer Sciences, Social Informatics and Telecommunications Engineering 2012

Nowadays, measurement systems only allow for a selective monitoring of environmental aspects for certain places, which will not be sufficient according to comprehensive monitoring tasks. In this context, a high-resolution and real-time environmental monitoring platform is highly beneficial in order to determine problematic areas and to plan appropriate counter-measures. This information is not only valuable for administrative purposes, but also for the general public, e.g. services that provide up-to-date information about healthy regions to live or routes to get from one point to another.

Sensor networks have improved and leveraged the environmental observation especially in rough, dangerous and preservable areas. In this context, also mobile or people-centric sensing becomes increasingly popular. This progress led to cheaper production, improved measuring techniques, and small sized measuring units. We believe that in the near future mobile devices, like smart phones, will be equipped with environmental sensing units and therefore this information will be available ubiquitously.

In this work we present a *Collaborative Sensing Platform for Eco Routing and Environmental Monitoring*. On a related note, the necessary information is collected by mobile as well as stationary sensing units and will be processed at the platform to preserve high resolution and up-to-date information about the current air quality with the focus on urban environments. Furthermore, the platform is capable to provide certain pollution-related services like *Eco Routing* and visualization. The first one should help to improve administrative monitoring and (traffic) planning tasks, whereas the latter one can be advantageous for pedestrians and cyclists to minimize the exposure to air pollution and find more healthy routes through the town.

The remainder of this paper is structured as follows. In Section 2, surveyed related work is outlined before we discuss our approach of the platform in Section 3 including the main requirements, the system architecture as well as the relevant components for *Eco Routing* and visualization. The evaluation of our platform is examined in Section 4. Section 5 concludes the paper and provides an outlook on future work.

2 Related Work

In this section, we will give an overview on related research and projects. Within the first part, platforms to obtain, manage and evaluate sensor data will be addressed. The second part consists of different technologies for modeling environmental pollution data. Finally, two bicycle routing projects will be introduced, which allow for specific routing options.

Eisenman et al. proposed a mobile sensing system, BikeNet [1], for mapping cyclist experience. In this project, sensors are directly embedded in bicycles and gather information about rides and according environmental information. The data is either stored offline and will be transferred to the platform later on or it will be transfered via so called Service Access Points (SAP) in real-time. The platform can be used to share cycling related data, like favorite routes, as well as data of more general interest, like pollution data. With regard to our system the proposed approach concentrates only on the pollution of the actual driven routes and therefore does not allow for high resolution monitoring capabilities.

CarTel [2] is a mobile sensor computing system, proposed by Hull et al., which was designed to collect, process, deliver and visualize data. The information is obtained from sensors located on cars. The gathered data is preprocessed on a so called CarTel node, before it is delivered to a central backbone for further analysis. The system provides a query-oriented programming interface and can handle large amounts of heterogeneous sensor data. The communication of the nodes primarily relies on opportunistic wireless connectivity to the Internet or to other nodes. CarTel applications use the collected data e.g. to optimize the travel distance of commuters or to localize areas with a high traffic density based on a low average speed. Nevertheless, the environmental information only addresses the emissions produced by the vehicle itself and not the pollutant concentration in the surroundings as proposed in our work.

SenseWeb [3], presented by Luo et al., is an open and scalable infrastructure for sharing and geocentric exploration of sensor data streams. It allows for indexing and caching of sensor data as well as processing spatio-temporal queries on real-time and historic data. Furthermore, aggregation and visualization of the obtained information is provided by a web-interface. In this work only stationary sensors or wireless sensor networks are utilized, whereas changes in position and density are not considered so far.

In the domain of statistics, there exist several approaches for data preparation and prediction which are often used in medical studies, e.g. the influence of air pollution on live expectation. In this context, several models already exist to describe phenomena like air pollution. Jerrett et al. compared five different classes of models in their work [4]: proximity, interpolation, regression, dispersion and meteorological models. These models were also subject to certain improvements and specializations within several publications [5,6,7,8,9], which will not be explained in more detail due to space limitations.

Cyclevancouver[2] is a a free web-based service that allows for bicycle route planning in the city of Vancouver (Canada). The objective is to encourage citizens in cycling. The service provides optimized routes especially for bikes and allows for additional options like low ascent, abundant vegetation, and low air pollution. However, the pollution-based route generation utilizes a statical model of the year 2003 that was only calculated once and therefore does not consider the current environmental situation.

Fahrradies[3] is also a web-based service which offers an interactive navigation system for cycle tours within the Oldenburger Münsterland (a region in North Rhine-Westphalia, Germany) [10]. The multifaceted routing options are very flexible and the navigation engine also allows for several intermediate stops (sights, shops, etc.). By using the *green route* option only the environmental surroundings in terms of the vegetation and development are considered.

---

[2] http://cyclevancouver.ubc.ca, last access 01.02.2012.

[3] http://www.fahrradies.net/, last access 01.02.2012.

3   Collaborative Sensing Platform

In this section, we will introduce the requirements, a collaborative platform for environmental monitoring and the provisioning of related services must fulfill. In this context the term *collaborative* refers to the collaboration of users in collecting measurements. Afterwards, we present the architecture and explain the main components in more detail, whereby the basis was the final thesis of Alexander Müller [11]. Finally, we outline a service called *Eco Routing*, which utilizes the obtained and preprocessed sensor data to allow for healthier and less polluted routes. The second service concentrates on the visualization of pollution data for different purposes.

3.1   Requirements

To enable high resolution measurements, not only stationary sensors will be considered, but also mobile ones. In order to use the provided data in a geospatial context, information about the location must be obtained. The location of stationary sensors is known by definition, whereas mobile sensor devices require additional positioning capabilities, e.g. GPS receiver. For environmental monitoring in the context of traffic and transportation, suitable sensors for NO_2, CO, CO_2, etc. should be used.

Scalability and robustness are also important requirements to preserve high resolution pollution information, wherefore large amounts of sensory data must be handled and processed. In this context, a viable storage and update structure is essential. Moreover, the provided information certainly is subject to wild fluctuations. These fluctuations do not only arise from the quality of different measuring devices, but also by the dimensions of space and time. During a single day, there will be more crowded areas leading to more measurements and a higher density as well as there will be certain areas that are less crowded, leading to sparseness of measurements. Therefore, suitable techniques are necessary to allow for an equalization of the data and provide appropriate information about the degree of pollution at a certain place and time.

Besides the information basis, also the service provisioning demands certain needs. The most important one is to map the pollution information onto the corresponding road segments as weights that can be used during the routing process. Additional information about the street network itself is essential, which should allow for routing and can be easily separated into the mentioned road segments.

3.2   Architecture

Before we explain the details of the proposed system to enable environmental monitoring including an *Eco Routing* and a visualization service, we will give an overview of the architecture and the involved components. We decided for a layered approach, which is illustrated in Figure 1.

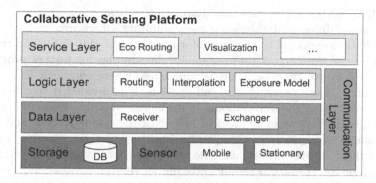

Fig. 1. System architecture overview

The bottom layer consists of two parts, namely sensor and storage. The sensor sector represents the mobile and stationary sensor devices, which gather information about the environment in form of pollution data as mentioned in the requirements section. The storage sector is responsible for storing the persistent information like e.g. the street network, results received by the upper layers, or as archive for historical purposes.

The data layer mainly receives the sensor data, adds appropriate timestamps, does minor plausibility checks and partitions the data samples according to their location before it is passed to the storage sector for further processing. Another task of this layer is to exchange data from the lower layer to the upper ones and vice versa. This becomes important when the platform is deployed in a distributed manner, whereby different communication channels can be used to exchange data with spatially separated components.

Within the logic layer, there are three main components, which are responsible for routing interpolation, and model generation. These three components will be explained in more detail later on. The vertical communication layer provides different communication standards to allow for an independent data exchange using various communication channels. Thereby, also a distributed interaction between the layers (storage, data, logic) is feasible.

The top layer represents services, which can be provided by the platform. Examples are a service, called *Eco Routing*, which allows for route calculations based on the current environmental situation in order to minimize the exposure to air pollution or a simple visualization or monitoring service, which can be used to receive up-to-date information about the pollution within a desired area.

Data Interpolation. In the first place, the sensor platform needs to collect and organize all sensor data received from the stationary and mobile sensors. This data includes:

- The GPS coordinates where the measuring took place
- The sensor type, e.g. information about the measuring error
- The measuring values of the air pollutant concentrations

The platform adds a server-side timestamp to the measuring samples on reception and uses a different database for each pollutant afterwards.

As the data is expected to be non-uniformly distributed in space and time, we require a special data structure to organize the samples for further processing. For that reason, we propose a spatial quadtree, where each cell has a maximum capacity. If the capacity of a cell exceeds, it is divided it into four child cells, each containing the same amount of measuring samples inside (0.5 quantile split). Figure 2a illustrates the logical layout of the quadtree cells and Figure 2b an example of a quadtree with quantile splitting method, showing a map of Bavaria with simulated sensor data. The measuring density in cities like Munich, Nuremberg or Regensburg is significantly higher than in rural areas.

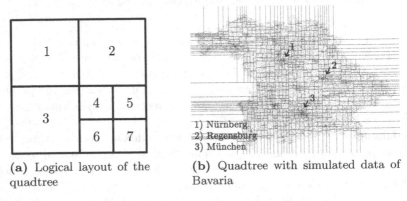

(a) Logical layout of the quadtree

(b) Quadtree with simulated data of Bavaria

Fig. 2. Quadtree data structure

For the interpolation part, we utilize a geostatistical interpolation technique called kriging [12,13], which models a spatial area as a multi-dimensional random process. To ensure real-time capabilities, we are only using the most recent measuring samples. As this method requires the underlying stochastic process to be intrinsically stationary, we use the quadtree cells as window for localized semivariogram estimations and a subsequent local kriging that is based on the nearest neighbor data samples. If a quadtree cell is spatially too large to assume stationarity, we try to subsequently split it using the same quantile based method as mentioned above. If this is not possible, e.g. there are not enough measuring samples inside, we try to use older measuring data. The semivariance of a local area is strongly influenced by geographic features as roads, landscape, buildings, etc., that do not change very quickly. So we are looking for a time slice, where the measuring density was high enough for a solid variogram estimation and use that data for the process. If this also fails, we use the whole cell for the semivariogram estimation and are forced to tolerate a small amount of non-stationarity instead of having no data at all for the according area.

As both, the variogram estimation and the kriging, require the data to be normally distributed, we need to transform the input data first by using a log-transformation. For the semivariogram estimation we use different variogram

models, that are put in a candidate list in the first place. One by one, we try to
fit them to the measuring data and if the fitting fails, the candidate gets removed
from the list. Finally, the candidate with the lowest residuals is selected as model
for the subsequent kriging interpolation. The candidate models are: (i) Spherical
model, (ii) Exponential model, (iii) Gaussian model, and (iv) Matern model with
M. Stein's parameterization.

We support two different types of kriging, each having a different objective.
The first approach uses localized block kriging to estimate a raster map of the air
pollutant concentration over a large area. This way we can gather information
not only about the street network, but also about the neighborhood, i.e. residual
areas, parks, etc. Using block kriging additionally smooths the interpolation
result, thus reducing the influence of potential outliers in the input data.

The second approach is to use a localized ordinary kriging to interpolate fixed
points on the street network, which are determined using a geospatial data set.
For each road we define a maximum segment length s. We split the road into
small segments with maximum length s by dividing the total length l of the road.
The segment count is given by $ceil(\frac{l}{s})$. We apply the ordinary kriging to the end
points of the road segments that have been determined by the previous step.
The second approach can only give information about the air pollution that is
present directly on the street network itself. It does not offer further information
about adjacent areas.

Both kriging types can account for known measuring errors by decomposing
the variogram's nugget into a model-based semivariance and an error-based one.
The interpolation result is then back-transformed to the original distribution.

Exposure Model. After the interpolation of the pollutant concentration, we
need to apply an exposure model to calculate an edge weight for the routing
graph for each road. We assume that the total exposure of a road user is basically
proportional to the retention time t and the concentration k on the road. Using
this simplifying assumption, we can calculate the total exposure E using

$$E = \int_0^T k(t)\, \mathrm{d}t$$

for a continuous function k or

$$E = \sum_i \Delta t_i \cdot k_i$$

for a discrete case, which can easily be derived from the interpolated data. The
process is shown in Figure 3. The exposure E is equal to the total area of the
gray surface.

The calculation of the total exposure E is split into two separate steps. We
interpret the road as a one-dimensional function from 0 (start point) to l (end
point), so we can define a function

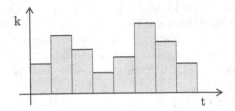

Fig. 3. Function $E_s(s)$ as discrete function

$$E_s : [0, l] \mapsto \mathbb{R} \ ,$$

which gives us the pollutant concentration for all points on the road. We can calculate E_s using the discrete data gathered by the interpolation step. With the raster interpolation, we can intersect the road with the raster cells to get an ordered list of road segments, each having a particular concentration value. In the point interpolation case, we already precomputed the road segments, which were used to determine the interpolation point coordinates. Using these points, we define new segments, each having one interpolation point in the center, which represents the concentration value for the whole segment. In doing so, we can interpret the segments and the associated pollutant concentration as a discrete representation of the function $E_s(s)$.

To incorporate the time into the model, we define a function

$$E_t : [0, T] \mapsto \mathbb{R} \ , E_t = E_s \circ s(t) \ ,$$

with T being the total driving time on that road and $s(t)$ giving the location of the vehicle at time t.

T and $s(t)$ are dependent of the velocity v on that specific road. In the simplest case, we can assume a constant speed v_{const}, that is either determined by an explicit speed limit or an implicit average speed for that kind of road. In some advanced cases, e.g. if we use floating-car data to estimate the actual speed, the velocity v can be given as coordinate-dependent function $v(s)$. To estimate E_t, we need to calculate the function $s(t)$, which can be derived from $v(s)$ by:

$$dt = \frac{ds}{v(s)} \quad \Rightarrow \quad t(s) = t_0 + \int_0^s \frac{d\bar{s}}{v(\bar{s})} \quad \Rightarrow \quad s(t) = t^{-1}(s)$$

And in the case of a constant velocity: $s(t) = v_{const} \cdot t$.

Eco Routing. The generated model, explained in the previous section, is the basis for the routing component. Thereby, we propose an approach that uses the exposure value calculated beforehand, which is static when the routing process starts. Hence, no dynamic edge weights will be used and in case of a long distance route, it should be recalculated after defined timeouts. But, as the routing service mainly addresses the inner-city traffic the occurrences of long distance travels is quite rare.

In general, routing requires a weighting function for the edges in order to determine the costs for a specific way. Classical routing strategies like the *shortest path* metric utilize the distance as weight, whereas the *fastest path* uses the allowed speed in combination with the distance to utilize the driving time as edge weight. Later on, these two approaches will be used for comparison during our evaluation in Section 4.3.

The *fastest eco path* metric is directly based on the exposure value E, as it is directly proportional to the driving time as it is a product of the driving time and the exposure to pollution for the according road segment. In case a faster vehicle drives along the road segment, the exposure will be lower than for a slower one.

The *shortest eco path* metric utilizes the calculated exposure value E only as basis for the edge weight. Hence, E will be normalized by dividing it by the driving time of the corresponding road segment. The interim value represents the average pollutant concentration, which is multiplied with the length l of the road segment to receive the final edge weight ω:

$$\omega = \frac{E \times l}{t}$$

Its advantage is that ω is independent of the driving time and therefore also independent of the vehicle's speed and the speed limit on that road.

Within our implementation, we utilize the A* routing algorithm, which was first mentioned by Hart et al. in [14]. The algorithm applies a heuristic that guesses the minimal distance between two vertices within a weighted graph $G = (V, E, \omega)$ consisting of vertices V and edges E and according edge weights ω. Vertices in a street network carry additional geographic information, which can be used to calculate the beeline as heuristic for A*, which also fulfills the triangle inequality in our case.

Visualization. The calculated pollutant concentration can be visualized according to the desired purpose. In the following, we present 3 methods, whereby the first two require raster data and the latter one operates on the pollutant concentration on the street network.

`Pollution data as Bitmap`: The visualization of raster information is often realized by using bitmaps. In our example, the interpolated values are mapped onto the value range $[0, N_{max}]$ with N_i as brightness value, resulting in a monochrome bitmap, where black represents areas with low pollution and white areas with a high pollution value. It is also possible to colorize the bitmap in a second step. Figure 4a illustrates monochrome and colorized pollution bitmaps using red for areas with a higher pollution and green for areas with a lower one.

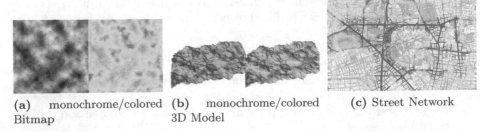

(a) monochrome/colored Bitmap (b) monochrome/colored 3D Model (c) Street Network

Fig. 4. Pollution visualization as Bitmap, 3D-Model, and Street Network

Pollution data as 3D Model: In comparison to the bitmap, the generation of a 3D model is much more computation-intensive. Thereby, the pollution value will be used as height information or the z-axis respectively. For rendering purposes a Point Cloud as well as a Mesh is reasonable. Optionally, a bitmap of the aerial perspective can be used as texture. Figure 4b shows the 3D model generated with the same pollution values as the bitmap before.

Pollution mapping on Street Network: In case the exposure values were calculated for the street network, also a direct mapping on the according road segments is possible. Thereby, the color information is directly derived from the current values. One option is to use the exposure value E and the other one is the utilization of the exposure function E_s. In the first case, the exposure value must be normalized onto the length of the road segment, otherwise longer road segments with an equal exposure value will have a higher (color-)value than shorter ones. For both cases, similar to the bitmap approach, the exposure values must be initially mapped onto an according (color-)value range. Figure 4c shows a colorized part of the street network close to the Olympic Park of Munich, whereby blue indicates the highest pollution and green the lowest.

4 Evaluation and Results

In this section, we present the evaluations conducted for our platform for routing and environmental monitoring. Therefore, we first describe how we obtained the sensor data needed for our evaluations, followed by the analysis of our interpolation methods and the *Eco Routing* approach.

4.1 Sensor Input

The proposed platform assumes that large-scale sensor information from distributed mobile sensors is available. Sensors could be integrated in user devices, such as smartphones, or mobile vehicles, such as cars or bicycles. However, since this kind of sensors are not yet available in large quantities, we had to generate realistic sensor input data for the following evaluations.

For this purpose, we used *Land-use regression* (LUR). Since LUR requires a broad data basis, which was not available in our case, we applied an adapted model [6] that integrates information about the street network, traffic density and absolute altitude. The street network information is extracted from *OpenStreetMap* project, the traffic density is estimated based on the type of road and the amount of lanes, and the absolute altitude was determined with the ASTER GDEM dataset[4].

The following regression model is used for generating synthetic NO_2 data:

$$\mu(NO_2) = \beta_0 + \beta_1 \cdot (15 \cdot TD_{0-40} + TD_{40-300}) + \beta_2 \cdot log_{10}(Alt)$$

TD_{n-m} denotes the traffic density in the area between n and m meters around the measuring point. The traffic density for a road segment is calculated by multiplying its length with a factor for the type of road. TD_{n-m} is the sum of densities for all road segments in the specified area. β_0-β_3 are weighting coefficients, which are mostly based on values used by Briggs et al. [7] in Huddersfield (UK), though β_1 was adapted to receive feasible results in our setting. For our simulations, we used the following values: $\beta_0 = 38.52$, $\beta_1 = 0.0003705$ and $\beta_2 = -5.673$.

4.2   Data Interpolation

As mentioned in Section 3.2, the proposed platform supports two interpolation methods: point and raster interpolation. We evaluated those methods by comparing the difference of exposure obtained with raster interpolation (exp_{raster}) and point interpolation (exp_{point}) for each street segment and determining the ratio $q = \frac{exp_{raster}}{exp_{point}}$. The analysis was conducted for various maximum distances of measuring points: 15, 25, 35, 45, 55 and 75 meters.

Table 1. Minimum, maximum, mean and standard deviation of q

Distance d [m]	Min	Max	μ	σ	Segment length [m]
15	0.654	1.558	0.9995	0.0120	12.03
25	0.657	1.558	0.9995	0.0120	18.09
35	0.659	1.558	0.9993	0.0122	23.28
45	0.662	1.558	0.9993	0.0123	27.91
55	0.662	1.558	0.9992	0.0125	31.92
75	0.660	1.558	0.9991	0.0130	35.47

The results for q are listed in Table 1. It can been seen that the mean value (μ) for q is in all case very close to the ideal value of 1.0, which shows that both interpolation methods perform almost similar. The standard deviation (σ) is also very low and ranges from 0.0120 (for $d = 15$) to 0.0130 (for $d = 75$), so that for 99% of the tested roads both values deviate less than 5%.

We also analyzed the performance of both approaches with a standard PC from the premium price segment. Even though the absolute numbers will be

---

[4] http://www.gdem.aster.ersdac.or.jp/, last access 01.02.2012.

different on other computers, the relative results indicate the performance of the approaches for different settings.

Figure 5a shows the time measurement results for point interpolation with varying distances between measuring points. It can be observed, that the duration rapidly decreases with greater separation of measuring points. In our setting, a more fine-grained interpolation than 45m should be avoided, as the run time sharply increases for lower distances.

The raster interpolation is only dependent on the raster size. Thus, we examined the time duration for this kind of interpolation with different raster sizes. In our simulations, we analyzed the run times for rasters with varying side length, ranging from 1 to 5 arcseconds. The results are illustrated in Figure 5b. It can be seen, that the required time is greatly reduced with greater side lengths. However, it does not scale in a quadratic way as one could imagine because of the quadratic scaling of the raster area, which mainly results from additional exposure calculation.

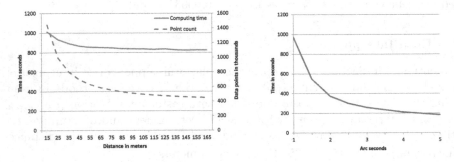

(a) Run time evaluation for point interpolation

(b) Run time evaluation for raster interpolation

Fig. 5. Performance analysis for point and raster interpolation

4.3   Eco Routing Comparison

In this section, we analyze the effects of the proposed *Eco Routing* on driving distance, driving time and exposure to air pollution exposure. We therefore compared the traditional routing algorithms *shortest path* and *fastest path* with the results of the corresponding *Eco Routing* algorithms.

Simulation Setup. To evaluate the different routing approaches, we simulated realistic air pollution with the above mentioned LUR-based approach for the simulation area (Munich, Germany) and compared the resulting routing decisions based on this data.

For the routing evaluation, we generated a large set of different routes. Therefore we selected a set of 55 starting and end points in Munich. All selected points represent some point of interest (e.g. metro station, popular square, etc.) and

(a) Selected starting and end points in simulation area

(b) *Shortest path* with / without *Eco Routing*

Fig. 6. Eco Routing simulation

are reachable by car and bicycle (e.g. no pedestrian area or highway). Further, all points are well connected to street network, i.e. routes in all directions are possible without long detours. The selected points are marked in Figure 6a.

For each of these starting points, routes to all other points were calculated, resulting in 2970 different starting point/end point combinations. Subsequently, 6 routes were determined for each of this combinations:

- Two routes determining the shortest path for cars, once with the traditional approach and once with our proposed *Eco Routing*. An example is shown in Figure 6b.
- Two routes determining the fastest path for cars, again without and with applied *Eco Routing*.
- Two routes for bicycles (using the fastest path). As we assume a constant speed of 16 km/h of cyclists, irrespectively from the chosen road segment, the shortest and fastest path are the same. Consequently, for the bicycle profile, we evaluated only the fastest path with and without *Eco Routing*.

In total this leads to 17820 calculated routes in our simulation.

Simulation Results. In order to compare the results from the traditional routing with our *Eco Routing*, we captured the relative difference for driving distance (len_q), driving time ($time_q$), and pollution exposure (exp_q) with

$$q := \frac{v_e - v_n}{v_n} = \frac{v_e}{v_n} - 1,$$

where v_n and v_e denote the resulting values from the traditional routing approach and the exposure-based approach respectively. As a result, negative values for q represent a decrease of the corresponding parameter when applying the proposed *Eco Routing*, positive values for q indicate a lower value for the traditional approach.

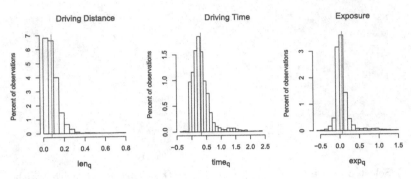

Fig. 7. Resulting relative difference for the *shortest path* approach for cars

In the setting of routing for cars with the *shortest path* approach, the traditional routing outperformed the *Eco Routing* approach. Figure 7 shows that the average of all three evaluated values increased. Remarkable is, that even the pollution exposure increased by 6.7%, which results from the fact, that the exposure is calculated as a product of time and pollutant concentration. Even though the average exposure might be reduced, the total exposure is increased due to a 32.2% increase of the average driving time.

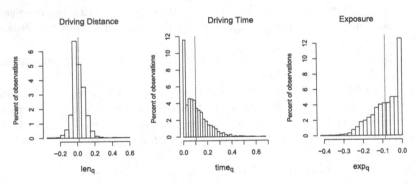

Fig. 8. Resulting relative difference for the *fastest path* approach for cars

The results of the *fastest path* simulations for cars (cf. Figure 8) show that the average pollution exposure could be significantly reduced. In our simulations, a decrease of 8.6% of pollution exposure could be observed when using the *Eco Routing* approach. However, the driving distance increased by 1.9% and the driving time increased by 10.3% in average. The latter is mainly the result of routes with a lower speed limit.

The simulations for the *fastest path* for bicycles led to the biggest decrease of pollution exposure (cf. 9). The average exposure was 12.6% lower with the proposed *Eco Routing* compared to the traditional approach. This was reached with only a moderate increase of driving distance and time, which both increased

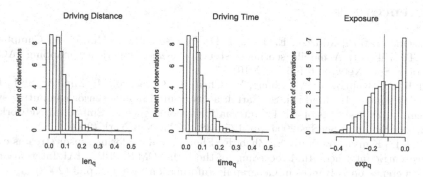

Fig. 9. Resulting relative difference for the *fastest path* approach for bicycles

by 7.1%. A more detailed examination shows that for 10% of the simulated routes the exposure was decreased by even 25%, and for 30% routes it still reduced the exposure at least by 17%.

5   Conclusion and Future Work

In this work, we presented our approach of a *Collaborative Sensing Platform for Eco Routing and Environmental Monitoring*. The proposed architecture enables a scalable and robust collection and processing of ubiquitous sensor data. The obtained information consists of traffic-related pollutants like NO_2, CO, CO_2 and allows for accurate and high resolution environmental monitoring in urban environments. Furthermore, we presented two services including their evaluation: *Eco Routing* and visualization. The former one calculates routes with the objective to minimize the exposure to harmful emissions. The results obtained by simulation indicate that exposure for cyclists was 12.6% lower compared to a traditional approach with a moderate increase by 7.1% of both the driving distance and the driving . For the latter service, we illustrated different techniques to visualize the pollution information for monitoring purposes. In this context, the interpolation process conduces as a basis, wherefore we carried out a performance analysis and presented the corresponding results.

As our sample measurements were generated using a *Land-use regression* model it would be beneficial to obtain more realistic sensor input data. Therefore, a prototypical low power sensor board connected to a smartphone will be applied in a coming field test. On the one hand, future work concentrates on the implementation of additional services like a dynamic pricing system considering current environmental situation. Based on this, a flexible adaptation of the city toll or of fees to rent bicycles can be realized in order to improve the air quality. On the other hand, the collection of data is also in the focus, as continuously taking and sending measurements is cost-intensive in several aspects (energy, traffic, money). Another future task is the performance enhancement of the system by the utilization of cloud computing technologies.

References

1. Eisenman, S.B., Miluzzo, E., Lane, N.D., Peterson, R.A., Ahn, G.-S., Campbell, A.T.: Bikenet: A mobile sensing system for cyclist experience mapping. ACM Trans. Sen. Netw. 6, 6:1–6:39 (2010)
2. Hull, B., Bychkovsky, V., Zhang, Y., Chen, K., Goraczko, M., Miu, A., Shih, E., Balakrishnan, H., Madden, S.: Cartel: a distributed mobile sensor computing system. In: Proceedings of the 4th International Conference on Embedded Networked Sensor Systems, SenSys 2006, pp. 125–138. ACM, New York (2006)
3. Luo, L., Kansal, A., Nath, S., Zhao, F.: Sharing and exploring sensor streams over geocentric interfaces. In: Proceedings of the 16th ACM SIGSPATIAL International Conference on Advances in Geographic Information Systems, p. 3 (2008)
4. Jerrett, M., Arain, A., Kanaroglou, P., Beckerman, B., Potoglou, D., Sahsuvaroglu, T., Morrison, J., Giovis, C.: A review and evaluation of intraurban air pollution exposure models. Journal of Exposure Analysis and Environmental Epidemiology 15(2), 185–204 (2004)
5. Henderson, S.B., Beckerman, B., Jerrett, M., Brauer, M.: Application of land use regression to estimate long-term concentrations of traffic-related nitrogen oxides and fine particulate matter. Environmental Science & Technology 41(7), 2422–2428 (2007)
6. Briggs, D.J., Collins, S., Elliott, P., Fischer, P., Kingham, S., Lebret, E., Pryl, K., van Reeuwijk, H., Smallbone, K., van der Veen, A.: Mapping urban air pollution using gis: a regression-based approach. International Journal of Geographical Information Science 11(7), 699–718 (1997)
7. Briggs, D.J., de Hoogh, C., Gulliver, J., Wills, J., Elliott, P., Kingham, S., Smallbone, K.: A regression-based method for mapping traffic-related air pollution: application and testing in four contrasting urban environments. Science of The Total Environment 253(1-3), 151–167 (2000)
8. Dabberdt, W.F., Ludwig, F.L., Johnson Jr., W.B.: Validation and applications of an urban diffusion model for vehicular pollutants. Atmospheric Environment (1967) 7(6), 603–618 (1973)
9. Nelson, L., Seaman: Meteorological modeling for air-quality assessments. Atmospheric Environment 34(12-14), 2231–2259 (2000)
10. Ehlers, M., Jung, S., Stroemer, K.: Design and implementation of a gis based bicycle routing system for the world wide web. International Archives of Photogrammetry Remote Sensing and Spatial Information Sciences 34(4), 425–429 (2002)
11. Müller, A.: Plattform für Echtzeit-Umweltdaten zur Navigation basierend auf ubiquitären Sensorinformationen (February 2012)
12. Krige, D.G.: A statistical approach to some basic mine valuation problems on the witwatersrand. Journal of the Chemical, Metallurgical and Mining Society of South Africa 52(6), 119–139 (1951)
13. Cressie, N.A.C.: Statistics for Spatial Data, rev. edn. Wiley series in probability and mathematical statistics. Wiley, New York (1993)
14. Hart, P., Nilsson, N., Raphael, B.: A Formal Basis for the Heuristic Determination of Minimum Cost Paths. IEEE Transactions on Systems Science and Cybernetics 4(2), 100–107 (1968)

Integrate WSN to the Web of Things by Using XMPP

Pin Nie and Jukka K. Nurminen

Aalto University, School of Science, Finland
{pin.nie,jukka.k.nurminen}@aalto.fi

Abstract. Wireless Sensor Network is a promising technology thanks to its numerous beneficial applications. The recent trend towards Web of Things leverages substantial web technologies and toolkits, which greatly simplify the chore of WSN application development. However, the complex web server and heavy HTTP communications impose difficulties on portability of WSN applications and node's resources management. In order to provide a lightweight web integration and uniform data representation, we propose to employ XMPP, an open standard formalized by IETF, to build instant messaging and presence service for wireless sensor nodes. In this paper, we develop a scalable and flexible XMPP sensor bot to integrate WSN into generic XMPP architecture. We also design two lightweight XMPP extensions for sensor node representation and task configuration. The efficient XML expression in our extension protocol can squeeze the payload into a single IEEE 802.15.4 packet and does not cause XMPP message fragmentation. Our solution works directly on MAC layer without the need of TCP/IP stack. Based on our sensor bot, we propose a novel application for product validation and customer behavior analysis with RFID/NFC technology on smartphones to demonstrate a new context-aware service.

Keywords: Wireless Sensor Network, XMPP, Instant Messaging and Presence Service, RFID/NFC Application.

1 Introduction

Humans have intrinsic limitations to observe the surrounding world in terms of the sensitivity of our natural perception of the environment, the variety of our senses and the persistent working capability. Moreover, we cannot directly perceive hazardous substances, such as chemicals and radiation. Therefore, we need sensors to enhance our perception, to perform persistent monitoring and to expand our awareness. Particularly, wireless sensor networks (WSN) [1] connect different sensor nodes over distance to fulfill various tasks in a cooperative manner without costly cable infrastructure.

Despite many benefits, WSN has not yet been widely employed in practice for multiple reasons. Firstly, there are numerous wireless technologies, such as IEEE 802.15.4/Zigbee, WirelessHART, Bluetooth Low Energy (BLE), Near

F. Martins, L. Lopes, and H. Paulino (Eds.): S-Cube 2012, LNICST 102, pp. 105–120, 2012.
© Institute for Computer Sciences, Social Informatics and Telecommunications Engineering 2012

Field Communication (NFC), Z-wave, Dash7 and WiFi. These protocols and standards do not work together and cause severe interoperability issue when bridging multiple WSN applications. Secondly, resource constraints on sensor nodes set up physical barriers to build reliable and long lifespan WSN systems. Thirdly, incompatible proprietary WSN software create information silos and hinder application development. Consequently, these difficulties prohibit cooperation between different WSN applications for seamless service creation. To solve this problem, web integration brings a mature platform for better flexibility and scalability. Evolved from the Internet of Things (IoT) [2], the Web of Things (WoT) [3] is proposed to simplify WSN development by leveraging embedded web server and HTTP communications. This architecture fits a few regular WSN applications with powerful sensor nodes. However, the complex web server and heavy HTTP communications impose difficulties on portability of WSN applications and node's resources management. Many limited sensor motes need ultra-lightweight web integration in just a few kilobytes memory [4]. Furthermore, HTTP does not support real-time communication which is required in safety-critical and/or highly interactive WSN applications.

eXtensible Messaging and Presence Protocol (XMPP) [5] [6] is an open standard, XML-based communication protocol for instant messaging and presence information. XMPP can offer uniform data representation and real-time service. XMPP has been tested in some WSN application scenarios [7] [8] to disseminate data. Therefore, we propose to extend XMPP architecture for WSN integration on the Web. In this paper, we develop an XMPP client called *sensor bot* to integrate sensor nodes into generic XMPP architecture. In contrast to the static customized messages sent by other XMPP/Jabber bots, our sensor bot automatically generates events derived from sensor data based on the processing rules, and then encapsulates events into XMPP messages for further distribution. In addition, the sensor bot also accepts incoming requests in predefined patterns from authorized contacts to create new tasks and/or update existing tasks. Moreover, we add feedback study on user rating from their responses for parameters optimization, such as sampling frequency and/or event notification preference. Essentially, we explore three generic programming components to build a scalable and flexible XMPP bot for more WSN application logics and better interaction with end users. Meanwhile, to improve XMPP functionality regarding WSN characteristics, we design two XMPP extensions for sensor node representation and task configuration. Since many tiny wireless sensor nodes do not support TCP/IP stack, we build our XMPP extensions directly on MAC lay and squeeze the payload into a single IEEE 802.15.4 packet. Hence, our protocol completely removes the dependence on transport layer by eliminating XML message fragmentation and data packet serialization.

Based on our WSN-enabled XMPP architecture, we propose a novel application for product validation and customer behavior analysis. We use RFID/NFC technology on smartphones to build a new context-aware service. We measured end-to-end latency to evaluate real-time performance on a preliminary prototype. The experimental result proves the feasibility to integrate WSN to the

Web via open XMPP network on the Internet. Thus, our XMPP architecture can be reused by other WSN applications.

The rest of the paper is organized as follows: Section 2 studies related works on WSN web integration and compares with our XMPP-based solution. Section 3 explains the primary elements in XMPP network and our development of sensor bot. Section 4 presents two XMPP extensions for WSN integration. Section 5 elaborates a novel application on smartphones with RFID/NFC technology. At last, Section 6 summarizes the paper.

2 Related Works

The Web of Things is an emerging architecture with the purpose to connect a variety of limited devices over the Internet, such as mobile phones and wireless sensor nodes. In [9] [10], a web mashup is proposed for embedded devices using tiny web server through either a gateway or direct integration on the SunSPOT [11] sensor node with a built-in HTTP engine. To expand data sharing, [12] presents a WSN integration into social networks via Restful Web API. An extensive study on the integration of sensors and social networks is provided in [13]. A heavier and more sophisticated framework is Sensor Web Enablement (SWE) [14]. Open Geospatial Consortium (OGC) [15] has developed comprehensive SWE standards to integrate sensor devices, measurements, information models and services into the Internet.

Inspired by these novel development, we identify two critical elements to boost WSN integration to the Internet. The first element is a uniform data representation to encapsulate information in a widely supported format. The second element is a ubiquitous service to distribute information in a widely supported mechanism. Driven by these two considerations, we select XMPP as the platform to develop WSN web integration. XMPP employs universal XML data model to encapsulate general content into message payload for instant messaging and presence service, which is a generic data exchange mechanism widely supported by many web services. As an open industrial standard, XMPP has fostered a big user community [1] and numerous interoperable free software [2].

Compared with HTTP-based web server architecture, XMPP has two advantages for WSN applications. Firstly, XMPP is a real-time protocol which can better serve safety-critical WSN applications, e.g., fire detection and health care. Secondly, XMPP client is a more lightweight program than a web server in typical Client/Server (C/S) model in which most application logics reside at server side. Thus, XMPP client can fit into embedded systems and mobile devices easier. Although XMPP has developed a lot of extensions, the core feature is rather straight-forward and simple with three basic XML stanzas. In contrast to static webpages and *pull* method in HTTP, XMPP supports dynamic data transportation with *push* method which is more suitable for asynchronous event

---

[1] Jabber Organization: http://www.jabber.org/

[2] XMPP Standards Foundation: http://xmpp.org/

notification in WSN applications. Concerning XML efficiency on resource consumption, several binary XML protocols are being standardized, such as Efficient XML Interchange (EXI) [16] [17] formalized by W3C. Binary XML compresses verbose XML content into more efficient format for storage and transmission. Notice that existing EXI products [18] [19] can achieve two orders of magnitude smaller binary XML messages with less than 2MB implementation.

XMPP/Jabber bot [20] is a client program sending customized replies in response to time trigger and/or incoming messages from others. Following publish/subscriber paradigm [21] [22], XMPP bot provides events notification service to other peers in the contact list. This feature saves considerable time and energy for WSN data dissemination from blind polling new data at a fixed interval. One micro-XMPP implementation [23] demonstrated the feasibility to enable XMPP on resource constrained wireless sensor nodes. In comparison, our XMPP sensor bot excels in flexibility of data processing and payload efficiency with two lightweight XMPP extensions tailored for WSN packet.

3   XMPP Architecture for Wireless Sensor Networks

Due to the limited battery and long lifetime requirement, WSN is featured with narrowband communication with low processing capabilities. In contrast to the fast hardware upgrade in many embedded systems, wireless sensor nodes are scaling up slowly for two reasons. Firstly, most of WSN applications mainly require primitive sensory data to enable situation awareness. WSN serves as an interface between cyberspace and physical world. Sensing is the fundamental objective rather than processing. Secondly, current battery technology cannot power small sensor nodes for long lifespan and rich processing at low cost, particularly in the applications where large data is sampled (e.g., vibration, image and acoustic signals). Hence, we choose three types of devices, namely smartphone, sink node and powerful sensor nodes as the gateway to host XMPP sensor bot. In this section, we firstly introduce the generic XMPP framework and then explain our development of WSN-connected XMPP sensor bot.

3.1   Generic XMPP Framework

The XMPP architecture as shown in figure 1 is composed of three components: XMPP client, XMPP server and gateway to foreign networks. The XMPP client is an I/O interface for text and multimedia rendering and sending. The XMPP server is responsible for connection management and message routing. The gateway bridges different networks by translating different protocols into XMPP and vice versa. Any two XMPP elements use TCP connection for XML streaming session. A single TCP connection can carry multiple sessions identified by their unique ids. The identifier for XMPP entity (a.k.a., JID) follows a URI pattern: *user@domain/resource*. Three basic XML stanzas are defined in XMPP as follows. XML stanza is the message payload exchanged over the XML stream.

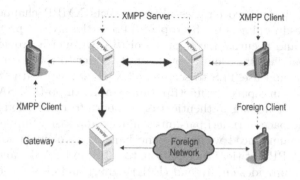

Fig. 1. The XMPP Architecture [24]

- <message/>: it is unicast carried out in store-and-forward mechanism through which one entity pushes information to another transferring messages between two endpoints.

```
<message from='niepin@aalto.fi' to='jukka@hut.fi'
        type='chat'>
  <body>Hello</body>
</message>
```

- <presence/>: it is broadcast executed in publish-subscribe mechanism through which multiple entities receive information about an entity to which they have subscribed, i.e., entity's availability.

```
<presence from='niepin@aalto.fi' xml:lang='en'>
  <show>online</show>
  <status>Working in the office</status>
</presence>
```

- <iq/> (Info/Query): it is a ping-pong interaction between two entities in request-response mechanism. This stanza can be used for service discovery and resource retrieval, such as file transfer and roster fetch. IQ interactions follow a common pattern of structured data exchange in the type of either get/result or set/result. An unique id is required to identify a transaction.

```
<iq type='get' from='niepin@aalto.fi' id='vq71f4nb'>
  <query xmlns='jabber:roster'/>
</iq>
```

XML stanzas must reside in the <stream/> block, which stands for an XML stream. All stream-level errors are unrecoverable and an <error/> stanza with description is sent by the detecting entity. Security constructs are specified in XMPP core protocol [5]. Two security protocols are employed to provide confidentiality, data integrity and entity authentication. The first protocol, Transport Layer Security (TLS), lies on the top of TCP connection and encrypts

XML streams between two entities. TLS protects XMPP channel from tampering and eavesdropping. On the top of TLS is the second protocol, Simple Authentication and Security Layer protocol (SASL), which provides a reliable mechanism to validate the identity of an entity. Prior to the SASL negotiation, XMPP clients should use TLS to secure the XML streams with "STARTTLS" extension in the namespace "urn:ietf:params:xml:ns:xmpp-tls". SASL defines a generic method for adding authentication support to connection-based protocols in the namespace "urn:ietf:par-ams:xml:ns:xmpp-sasl". Supported security mechanisms are announced within a <mechanisms/> element for negotiation between two XMPP entities. In addition to TLS and SASL, another XMPP specification [25] provides end-to-end (E2E) signing and object encryption.

Among plentiful XMPP extensions, Multi-User Chat (MUC) [26] is an important specification to enable many-to-many communication. This feature has high value to interconnect different WSN applications and share data for context-aware information processing. MUC is also beneficial to link up multiple WSNs at different locations for a bigger scale WSN application of the same interest or subject. All we need to do is to create a chat room, give a certain subject and invite others to join. A participant could be a WSN-enabled XMPP client publishing data and human users observing events at runtime.

3.2 XMPP Sensor Bot

From the generic XMPP framework above, we can leverage three key features for WSN applications as below:

1. Interactive communication: XMPP uses <presence/> stanza to indicate availability of an entity and <message/> stanza to exchange text messages in both online and offline cases. The presence status suits duty-cycled WSN applications in which sensor nodes are scheduled to sleep periodically for energy preservation. Whenever the node(s) wakes up and transmits data, associated contacts in the group will receive its updated information. Meanwhile, users can send queries or commands to get latest data or reconfigure tasks if authorized. When sensor nodes return to sleep, all offline messages will be cached in XMPP server till the next awake period.
2. Service discovery: based on the <iq/> stanza, an XMPP extension, XEP-0030 [27] specifies service discovery process for XMPP entities. We can define a JID URI pattern "wsn_name@domain/sensor_node_id/sensor" for resource binding on a specific WSN within the given domain. This URI allows one WSN operator to have multiple WSN applications which are identified by their names. Each WSN application may consist of many nodes and each node may equip several sensors for different measurements. The hierarchical structure provides flexible and scalable WSN resource binding.
3. Group chat: XMPP extension MUC offers rendezvous point for multiple WSN operators to share their real-time data in a common interest. This feature not only creates context-rich WSN application, it also covers a wide physical area. Based on a list of MUC room subjects, users can choose their favorite WSN application and observe a specific environment or monitor an interesting object.

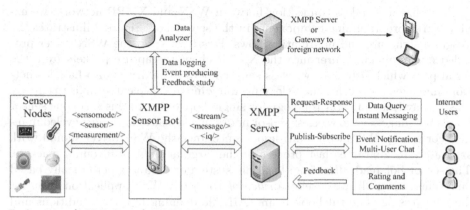

Environment → XML Data → XMPP Messages → Services and Applications → Value and Benefits

Fig. 2. WSN-enabled XMPP Architecture

In order to integrate WSN into generic XMPP architecture with minimal cost, we develop an XMPP sensor bot. Figure 2 illustrates the overall system from the sensor node to the Internet user. We create a small XMPP client [3] to collect data from sensor nodes and send customized messages to remote XMPP entities. A data analyzer is built into the sensor bot for data logging, event producing and feedback study. All other XMPP entities remain the intact without any change. Information flows smoothly from physical world to the Web via XMPP networks. Thus, our solution simplifies WSN application and service creation.

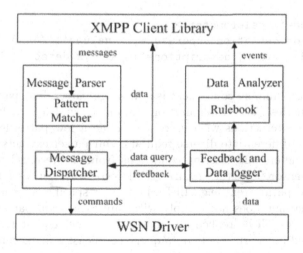

Fig. 3. Program structure of the sensor bot

---

[3] Smack API: http://www.igniterealtime.org/projects/smack/

To realize a seamless connection between WSN and XMPP network, we explore three programming components in the sensor bot. Figure 3 illustrates the program structure and information flows. Firstly, we add the WSN driver provided by the manufacturer into the XMPP client to capture packets from the serial port which listens to wireless sensor nodes. The driver also handles network management, such as node leaving and joining. A typical setup is to employ master-slave mode in star topology for single-hop WSN. In this case, one sensor bot can manage several wireless sensor nodes in a synchronized duty-cycled manner. For a large WSN with multi-hop connections, the WSN driver should work separately to guarantee fast process of the large amount of incoming packets. The sensor bot reads data from local file system or database periodically based on a timer. Secondly, we develop a *rulebook* to specify WSN application logics for data processing. The rulebook is an XML file defining filtering conditions and publishing events for each type of sensor. Triggered by new receiving packets from the WSN or a local timer, the data analyzer executes these rules to publish events to the subscribed peers in the contact list through XMPP <message/> stanza in multicast method. The following example shows two types of sensor with different filters and consequent events. The first temperature sensor applies threshold filter and generates alarm if the value exceeds 40 degrees. The second accelerometer sensor applies deviation filter on vibration measurement and detects impact during motion monitoring. The rulebook is flexible and scalable to contain more data aggregation techniques [28] [29].

```
<sensor type='temperature'>
    <filter id='threshold' operator='>' value='40'/>
    <event id='alarm' description='temperature is too high.'/>
</sensor>

<sensor type='accelerometer'>
    <filter id='deviation' operator='>' value='1'/>
    <event id='motion' description='impact is detected.'/>
</sensor>
```

The third programming component is a message parser. The parser reads incoming data queries, tasks configuration commands and feedback ratings. To support flexible interactions with users, we define a few patterns using regular expression to differentiate diverse requests. Our parser extracts sensor type, command parameters and filtering conditions from the received message. Accordingly, the sensor bot either returns the latest measurement for data query or set command parameters (e.g., LEDs blink in a specific color and order). If an incoming message does not comply with any expression patterns, a list of allowable patterns will be replied automatically to the requester. Unlike event multicast defined in the rulebook, data query is unicast and executed only once in request-response mechanism. On the one hand, the event notification is used in routine monitoring to avoid overwhelming raw data and to highlight special status and/or changes. On the other hand, real-time data query handles random and dynamic situations, such as customer service.

Tasks configuration and parameters optimization are important issues in WSN application development. Thus, we add feedback study to rate users satisfaction. Our message parser will prompt users to "like" or "dislike" the response. Subsequent comments will be recorded as reasons for feedback study later. By counting the number of "like" and "dislike", the WSN operator can evaluate the popularity and quality of his applications and services for further improvement. Moreover, our sensor bot supports an open control feedback loop based on the user's presence information. By reading the subscriber's presence, the sensor bot decides whether it should send the data right now or postpone when the remote user is busy or not available.

4   XMPP Extensions for Sensor Networks

In the program structure of our XMPP sensor bot above, WSN driver is a potential bottleneck, because sensor nodes do not use XML format to encapsulate their data. The sensor bot has to parse every packet and format into XML element for XMPP messages. As a result, packet transformation may exhaust the sensor bot when dealing with heavy network traffic. Furthermore, low level packet conversion also hinders XML parser development when porting a WSN application to another XMPP sensor bot. However, none of existing XMPP extensions suit WSN due to their huge resource consumption in complex signaling and verbose expression. To solve this problem, we propose two lightweight XMPP extensions to encapsulate XML format payload into a single IEEE 802.15.4 packet. Our XMPP extensions cover two fundamental functions in all WSN applications, namely node representation and task configuration.

4.1   Node Representation

There are two common attributes for every sensor node: capability and measurement. Capability specifies the equipped hardware sensors and the supported precision and format. One sensor node may have several sensors (e.g., temperature, light, accelerometer, barometer) onboard. Measurement delivers data from the hardware. One temperature sensor gives float value in degree centigrade. Measurement may also include data point of embedded software algorithms. One sensor node can calculate dew point based on temperature and humidity at a given altitude. With these two common attributes, a WSN application supports service discovery and data provision.

To represent a sensor node in a concise profile, we design a new XML stanza <sensornode/> with two child elements: <sensor/> and <measurement/>. As aforementioned, these two elements list equipped sensors and supported measurements onboard. In addition, a <sensornode/> has one basic attribute, 'id' and three optional attributes, 'type', 'location' and 'time'. The 'id' attribute identifies the node in WSN and can be appended in XMPP entity URI for resource binding. The 'type' attribute is a tag for grouping or classification depending on the WSN application. The 'location' attribute gives positional context information. The 'time' attribute gives measurement a time stamp and may also be used to update node's presence status. The following example shows two <sensornode/> stanzas. The first stanza is for service discovery at the initial stage and the second stanza is a regular data report.

```
<sensornode id='node_1' type='fire_detector'
            location='office' time='YY:MM:DD-HH:MM:SS'>
  <sensor type='temperature' unit='celsius'/>
  <sensor type='light' unit='lux'/>
</sensornode>

<sensornode id='node_1'>
  <measurement type='temperature' value='25'/>
  <measurement type='light' value='100'/>
</sensornode>
```

Once the initial stage is completed, we reduce syntax verbosity for subsequent measurement reports by removing quotation marks and using abbreviated letter(s) of every XML element and attribute, such as 'sn' for 'sensornode' and 'y' for 'type' as shown below. We also encode measurement types in one byte which provides 255 unique options, enough for all possible values. Assume node's id and every letter take one byte and every numeric value takes four bytes (floating point), the abbreviated expression in the following example saves up to 64% space compared with the previous version. Assume employing the IEEE 802.15.4 radio standard which allows maximum 102 bytes for payload per frame, a single packet can take up to 5 measurements including location and time attributes in a row. We use a flag 'wsn=true' in the sensor bot to indicate this compact payload strategy in XML format for efficient wireless transmission. In the case of big chunk of data set, such as vibration amplitudes and acoustic signals, only the first and the last packet use XML format to mark the start and the end of packet streaming. The rest packets in the middle contain only values for serialization.

```
<sn i=node_1>
  <m y=t v=25/>
  <m y=l v=100/>
</sn>
```

4.2   Task Configuration

The second extension uses 'get/set' methods in XMPP <iq/> stanza to execute task configuration. We use get/result transaction to fetch configurable parameters on a sensor node and set/result transaction to update parameters' value. Since all parameters belong to either hardware sensors or software algorithms, we embed a new XML element <param/> into <sensor/> and <measurement/> elements. In the following example, a sensor node returns sampling frequency for accelerometer sensor and two embedded calculations for light measurement, average value and threshold filter.

```
<iq type='get' from='niepin@aalto.fi'
    to='node_1' id='info_1'>
  <query xmlns='http://aalto.fi/wsn#parameters'/>
</iq>

<iq type='result' from='node_1'
    to='niepin@aalto.fi' id='info_1'>
  <query xmlns='http://aalto.fi/wsn#parameters'>
    <sensor type='accelerometer'>
        <param name='frequency' unit='Hz'/>
    </sensor>
    <measurement type='light'>
        <param name='average'/>
        <param name='threshold'/>
    </measurement>
  </query>
</iq>
```

When setting parameters, our extension allows independent use of <param/> to activate or deactivate a sensor or an embedded algorithm. In the following example, we deactivate accelerometer and set up a threshold filter for light sensor. Successful result or error response is replied from the sensor node. Note that error message uses predefined code to indicate possible reason.

```
<iq type='set' from='niepin@aalto.fi'
    to='node_1' id='config_1'>
  <query xmlns='http://aalto.fi/wsn#parameters'>
    <param name='accelerometer' value='false'/>
    <measurement type='light'>
        <param name='threshold' operator='>' value='100'/>
    </measurement>
  </query>
</iq>
```

If success case

```
<iq type='result' from='node_1'
    to='niepin@aalto.fi' id='config_1'/>
```

If error case (404 parameter not found)

```
<iq type='error' code='404' from='node_1'
    to='niepin@aalto.fi' id='config_1'/>
```

Like the previous extension for node representation, the extension for task configuration also supports abbreviated expression. Given a certain namespace in a moderate size of WSN (nodes number< 255), we assume all id and type values in XML elements can be encoded in one byte letters and numeric values take four bytes (floating point). Then, the set/result example above can be squeezed into a single IEEE 802.15.4 radio packet as below:

```
<iq y=s f=niepin@aalto.fi t=node_1 i=config_1>
  <q x=http://aalto.fi/wsn#parameters>
    <p n=a v=f/>
    <m y=l>
        <p n=t o=> v=100/>
    </m>
  </q>
</iq>
```

If success case

```
<iq y=result i=config_1 f=node_1 t=niepin@aalto.fi/>
```

If error case (404 parameter not found)

```
<iq y=e c=404 i=config_1 f=node_1 t=niepin@aalto.fi/>
```

Our XMPP extensions introduce a few new XML elements tailored for WSN and reuse XMPP core specification. The abbreviated expression presents an efficient XML format for wireless transmission. In a typical WSN standard IEEE 802.15.4/Zigbee [4], no packet fragmentation is needed. Therefore, our solution achieves good scalability and interoperability by applying uniform XML format at little cost of payload redundancy.

5 RFID/NFC Application for Product Validation and Customer Behavior Analysis

Fake products and expired food are two big problems in many countries. How to validate products or food in a short time is a challenge for three reasons. Firstly, there are numerous products and food in the world or just in a local grocery store. Centralized method is impossible to accommodate all products and to keep up-to-date information. Secondly, product validation service has to be ubiquitous for easy access. Meanwhile, this service should be low cost to motivate people for daily use. Thirdly, the whole validation process must be quick and responsive to massive requests.

Driven by three concerns above, we propose to use smartphones and our XMPP sensor bot to build a fast product validation service with RFID/NFC technology. Correspondingly, there are three advantages in our solution. Firstly, XMPP provides an open peer-to-peer architecture for direct and instant communication between producers and consumers through instant messaging and presence service. Secondly, smartphone is becoming a ubiquitous mobile computing device in people's daily life. High speed wireless technologies, such as Wifi and 3G, support wide connectivity via smartphones. Furthermore, equipped with GPS receiver, smartphones can provide location and time context information for accurate data processing. Thirdly, RFID/NFC technology on smartphones

---

[4] Zigbee Alliance: http://www.zigbee.org/

enables touch-and-see quick object identification. RFID tag is a cheap passive circuit which can be attached on any solid product.

We develop an XMPP sensor bot on a smartphone which equips NFC chip to read RFID tag on the product. RFID tag should contain at least two types of data, the product's series number and the JID of the producer's customer service. Meanwhile, the sensor bot also reads current location (e.g., GPS coordinate) and the time. By encapsulating all these information in an XMPP <iq/> stanza, the sensor bot sends a product validation request to the remote producer and displays the result on the smartphone. In this way, the customer can easily validate a product or food at real-time without typing or searching anything.

Similar to the barcode used in warehouse products management, the validation request consists of four key elements: <product ID, timestamp, location, dealer's signature>. By checking the time and the location attribute, the producer can validate if the product/food has expired or not. Fake products with invalid IDs or obsolete valid IDs will not pass the validation because a product's series number is unique in a specific area during a certain period. The dealer's signature provides basic security feature with a shared secret which can be used to hash the whole validation request. The producer may also add promotion information in the response for advertisement. In the following example, the producer confirms the valid product and also replies discount information. The dealer's signature is omitted for simplicity.

```
<iq type='get' id='product_series_number'
    time='YY:MM:DD-HH:MM:SS' location='City_District_X'
    from='customer@example.com' to='producer@company.com'>
  <query xmlns='product#info'/>
</iq>

<iq type='result' id='product_series_number'
    from='producer@company.com' to='customer@example.com'>
  <query xmlns='product#info'>
    <item name='product_name' type='wine' value='true'/>
    <item name='promotion' type='discount' value='promo_code'/>
  </query>
</iq>
```

We implemented a preliminary testbed to evaluate the feasibility and real-time performance. Our testbed uses the smartphone Nokia C7 to read RFID tag and forward the data over 3G/WiFi to a proxy which runs a script to encapsulate incoming data in XMPP messages and then send to the Gtalk client on another smartphone and a Jabber client on a laptop. Figure 4 illustrates the network structure of our testbed. We measured the end-to-end latency. It takes about three seconds from touching the RFID tag with one smartphone till the XMPP message appearing on the other smartphone and the laptop. A round-trip time (RTT) may take six seconds. The major latency comes from the middle box, a HTTP-XMPP proxy which performs protocols translation for two-way communication. Our next step is to remove this extra proxy and implement peer-to-peer XMPP messaging between the smartphone and a Jabber client. The native XMPP communication will be much faster without any middle box. In addition, we also notice that 3G connection adds longer delay than WiFi connection due to the extra signaling overhead with the base station.

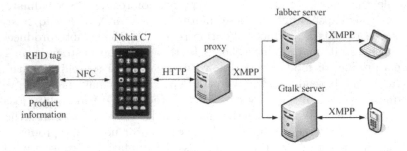

Fig. 4. Product validation with smartphone based on XMPP

In order to handle large number of request at low cost, the producer can register a JID on public XMPP servers and modify an open-source XMPP client [30] to automate product validation with a backend product database. Moreover, the producer can use this service to study customer behavior by counting the number of requests and exploring context information. This application can provide insights to three important questions that many producers concern:

1. Popularity: what is the total number of received requests during a certain period?
2. Customer distribution: at what time and in which place do customers buy this product?
3. Genuine-to-counterfeit ratio: how many fake products exist in the market?

6   Discussion and Conclusion

On highly constrained sensor nodes which cannot afford resources for TCP/IP stack, our lightweight XMPP extensions work directly on MAC layer for efficient XML encapsulation and transmission. More powerful devices can host our XMPP sensor bot, such as a smartphones, sink node connected with laptop or PC and high-end wireless sensor nodes (e.g., SunSPOT and Imote2-linux). Recently, IPv6 is gaining increasing adoption by many embedded devices and sensor nodes [31]. A number of efficient and reliable transport protocols have been also proposed on the lower layers for WSN [32]. Thus, our XMPP sensor bot will likely be able to run on low-end wireless sensor nodes in the future. Consequently, machine-to-machine communication will connect multiple XMPP sensor bots together to create heterogenous WSN applications at larger scale.

In this paper, we propose to use XMPP to integrate WSN to the Web of Things. Compared with another architecture based on web server and HTTP communications, our XMPP sensor bot can achieve real-time performance with small program footprint. We design two XMPP extensions for sensor node representation and task configuration. By reusing existing XMPP standards, our solution makes it easier to integrate WSN on generic XMPP architecture. At last, we propose a novel application for product validation with RFID/NFC technology to demonstrate the feasibility of our solution.

Acknowledgements. This research is partly funded by TEKES in the Internet of Things programme of TIVIT (Finnish Strategic Centre for Science, Technology and Innovation in the field of ICT).

References

1. Kuorilehto, M., Hännikäinen, M., Hämäläinen, T.D.: A survey of application distribution in wireless sensor networks. EURASIP Journal on Wireless Communications and Networking 2005, 774–788 (2005)
2. Atzori, L., Iera, A., Morabito, G.: The internet of things: A survey. Computer Networks 54 (October 2010)
3. Guinard, D., Trifa, V., Mattern, F., Wilde, E.: From the Internet of Things to the Web of Things: Resource Oriented Architecture and Best Practices. Springer (2011)
4. Mottola, L., Picco, G.P.: Programming wireless sensor networks: Fundamental concepts and state of the art. ACM Computer Survey 43 (April 2011)
5. Saint-Andre, P.: RFC 6120 Extensible Messaging and Presence Protocol (XMPP): Core (2011)
6. Saint-Andre, P.: RFC 6121 Extensible Messaging and Presence Protocol (XMPP): Instant Messaging and Presence (2011)
7. Goncalves, J., Ferreira, L.L., Chen, J., Pacheco, F.: Real-Time Data Dissemination for Wireless Sensor Networks using XMPP. Polytechnic Institute of Porto, Tech. Rep. (2009)
8. Hornsby, A., Belimpasakis, P., Defee, I.: XMPP-based wireless sensor network and its integration into the extended home environment. In: IEEE 13th International Symposium on Consumer Electronics, ISCE (2009)
9. Guinard, D., Trifa, V.: Towards the web of things: Web mashups for embedded devices. In: International World Wide Web Conference, Workshop on Mashups, Enterprise Mashups and Lightweight Composition on the Web, MEM 2009 (2009)
10. Guinard, D., Trifa, V., Wilde, E.: A resource oriented architecture for the web of things. In: Internet of Things, IOT (2010)
11. Sun small programmable object technology (sun spot) theory of operation, Tech. Rep. (2007)
12. Guinard, D., Fischer, M., Trifa, V.: Sharing using social networks in a composable web of things. In: 2010 8th IEEE International Conference on Pervasive Computing and Communications Workshops, PERCOM (2010)
13. Aggarwal, C.C., Abdelzaher, T.: Integrating sensors and social networks. In: Social Network Data Analytics, pp. 379–412. Springer, Heidelberg (2011)
14. Bröring, A., Echterhoff, J., Jirka, S., Simonis, I., Everding, T., Stasch, C., Liang, S., Lemmens, R.: New Generation Sensor Web Enablement. Sensors 11 (2011)

15. Botts, M., Percivall, G., Reed, C., Davidson, J.: OGC Sensor Web Enablement: Overview and High Level Architecture (2007)
16. Schneider, J., Kamiya, T.: Efficient XML Interchange (EXI) Format 1.0 (2011)
17. Cokus, M., Vogelheim, D.: Efficient XML Interchange (EXI) Best Practices (2007)
18. Peintner, D.: EXIficient: an open source implementation of the W3C Efficient XML Interchange (EXI) format specification in Java (2011), http://exificient.sourceforge.net/
19. Inc., A.: Efficient XML (2011), http://www.agiledelta.com/product_efx.html
20. Google wave bots (2011), http://googlewavebots.info/wiki/
21. Albano, M., Chessa, S.: Publish/subscribe in wireless sensor networks based on data centric storage. In: Proceedings of the 1st International Workshop on Context-Aware Middleware and Services: Affiliated with the 4th International Conference on Communication System Software and Middleware, COMSWARE 2009 (2009)
22. Millard, P., Saint-Andre, P., Meijer, R.: XEP-0060: Publish-Subscribe (2010)
23. Hornsby, A., Bail, E.: μXMPP: Lightweight implementation for low power operating system Contiki. In: ICUMT 2009 International Conference on Ultra Modern Telecommunications and Workshops (2009)
24. Laukkanen, M.: Extensible Messaging and Presence Protocol (XMPP). University of Helsinki, Department of Computer Science, Tech. Rep. (2004)
25. Saint-Andre, P.: RFC 3923 End-to-End Signing and Object Encryption for the Extensible Messaging and Presence Protocol, XMPP (2004)
26. Saint-Andre, P.: XEP-0045: Multi-User Chat (2008)
27. Hildebrand, J., Millard, P., Eatmon, R., Saint-Andre, P.: XEP-0030: Service Discovery (2008)
28. Rajagopalan, R., Varshney, P.K.: Data aggregation techniques in sensor networks: A survey. IEEE Communications Surveys and Tutorials (2006)
29. Nakamura, E.F., Loureiro, A.A.F., Frery, A.C.: Information fusion for wireless sensor networks: Methods, models, and classifications. ACM Computing Surveys 39(3) (2007)
30. Use XMPP to create your own google talk client (2010), http://web.sarathlakshman.com/Articles/XMPP.pdf
31. Montenegro, G., Kushalnagar, N., Hui, J., Culler, D.: RFC 4944 Transmission of IPv6 Packets over IEEE 802.15.4 Networks (2007)
32. Ayadi, A.: Energy-efficient and reliable transport protocols for wireless sensor networks: State-of-art. Wireless Sensor Network (March 2011)

Challenges and Opportunities for Embedded Computing in Retail Environments

Kunal Mankodiya, Rajeev Gandhi, and Priya Narasimhan

Department of Electrical and Computer Engineering, Carnegie Mellon University,
Pittsburgh, PA, USA
kunalm@cmu.edu, rgandhi@ece.cmu.edu, priya@cs.cmu.edu

Abstract. In the retail industry, real-time product location tends to be a multi-million-dollar problem because of seasonal restocking, varying store layouts, personnel training, diversity of products, etc. Stores maintain planograms, which are detailed product-level maps of the store layout. Unfortunately, these planograms are obsolete by the time that they are constructed (because it takes weeks to get them right), thereby significantly diminishing their value to the store staff, to consumers, and to product manufacturers/suppliers. The AndyVision project at Carnegie Mellon focuses on the fundamental problem of real-time planogram construction and planogram integrity. This problem, if solved correctly, has the potential to transform the retail industry, both in the back-office operations and in the front-of-the-store consumer experience.

Keywords: Retail operations, planogram, retail technology.

1 Introduction

For decades, the retail industry has been slow to adopt technology and has focused on fairly traditional technology, such as point-of-sale terminals, barcode scanners, loyalty cards and automated checkout. However, with an urgency to remain competitive in the face of the continual rise of online retail, brick-and-mortar stores want to make their environments more experiential and less staid. Increasingly, retailers are shifting from pure product lines towards ecosystems (McKinsey, 2000) of inter-related, distributed products, services and real-time information, including cross-channel (mobile, web, store, etc.) interactions. There is a push towards *making retail more personalized for the connected shopper in the connected store*, with customer behavioral data, social relationships and in-store/cross-store technology becoming key assets for retailers. Embedded computing has the power to transform the retail landscape. The *stakeholders here are both the shoppers and retail operations* in the back-office.

Digital retail technology can be used to meet shoppers' demands for efficiency, speed, improved customer service, as well as their desire to be more informed and socially connected, while simultaneously addressing the retail industry's needs for increased growth, increased basket-size ($$ of revenue per customer visit), reduced

F. Martins, L. Lopes, and H. Paulino (Eds.): S-Cube 2012, LNICST 102, pp. 121–136, 2012.
© Institute for Computer Sciences, Social Informatics and Telecommunications Engineering 2012

store staffing, improved back-office operations, and increased stickiness i.e., breeding customer loyalty by differentiating themselves from other retail stores. For example, retail stores have long wanted to improve customer service without having to add payroll costs through more store staff. This can be made possible through the strategic use of embedded computing at multiple points in the retail ecosystem—networked digital signs to detect shoppers' interest and engage them in purchases, in-store localization to determine the precise in-store locations of shoppers and products, sensors to track in-store traffic patterns, crowd-sourced analytics (from shopping history, social networks and other stores), mobile devices in the hands of shoppers to allow them to obtain information in real-time, robots to interact with customers and assist them at the point of purchase, etc.

"The holy grail of retailing – being able to offer the right product in the right place at the right time for the right price – remains frustratingly elusive" (Harvard Business Review, July-August 2000) still conveys the complete message of most retailers' day-to-day challenges. The retail industry is one of the most dynamic and influential industries in global economy. In the U.S., the retail industry represents about 40% of the Gross Domestic Product (GDP) and is the largest employer (Fisher and Raman, 2001). The intensified competition that the retail industry faces with the emergence of increasing numbers of new players in both the local and global markets has forced retailers to critically examine and redesign both their operations and marketing strategies (Perdikaki, 2009). To remain competitive, many retailers have differentiated themselves by designing enhanced in-store shopper experiences and other strategies to distinguish themselves from their competitors. Moreover, retailers must constantly strive for excellence in operations; extremely narrow profit margins leave little room for waste and inefficiency. Such practices have provided a fertile ground and new context for research as well as technological interventions in the retail-operations space.

In this paper, we present the current problems of retail operations and also propose a technological model as a solution.

2   Retail Operations

Every time that a shopper enters a store, his/her shopping experience has, in fact, been extensively planned, from the items that he/she sees immediately for sale at the entrance, all the way to the layout and design of the entire store. The decisions of planning and stocking are orchestrated by the retail-operations staff, who are closely concerned with the day-to-day functioning, operations and maintenance of the store. The retail-operations staff typically include sales clerks, check-out cashiers, supplier managers, sales people and floor managers; these roles are largely present in all stores in some form or the other, including both small stores with only a handful of workers and large chain stores with hundreds of employees. Figure 1 shows the different activities and aspects of retail operations.

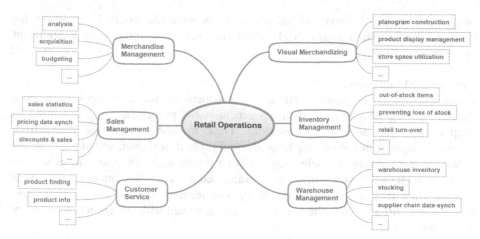

Fig. 1. Different aspects of retail operations

Merchandise Management

The primary function of any retail store is to sell merchandise; therefore, merchandise management is a key activity of the retail operations. Merchandise management involves in-depth analysis, planning, acquisition, handling and budgeting as related to merchandise (Gaffney, 2000):

- Retailers must correctly analyze their customers and their needs.
- Retailers must plan for proactively ordering merchandise to be sold in the future.
- Retails must procure merchandise from the appropriate suppliers and distributors.
- Retailers need to handle different categories of merchandise, schedule the appropriate time for promotion and display, as well as the right absolute locations (front of the store, aisle end-caps, back of the store, etc.) and the relative locations (i.e., next to which other kinds of products).
- Budgeting is an essential part of merchandise management on multiple levels, including restocking, redoing the store layout, seasonal needs, etc.
- Retailers must be able to perform inventory counts, on a sufficiently regular basis and with sufficient accuracy, in order to provide store audits, to ensure supplier compliance, to enable efficient restocking, etc. This is particularly the case for stores that are subject to restocking and planogram changes due to seasonal variations.

Visual Merchandizing

Visual merchandising is the activity and profession of developing floor plans and three-dimensional displays in order to maximize sales (Wetz, 1995). Both goods and services can be displayed to highlight their features and benefits. The purpose of such visual merchandising is to attract, engage and motivate the customer towards making a purchase. A "planogram" is often widely used in today's retail stores as a means of visual merchandising. A planogram is a detailed visual map of the store's products that indicates the placement (down to the aisle, shelf and height of the shelf) of products in order to maximize sales and to provide the optimal placement for suppliers (Oxford Dictionaries, 2011). Real-time planogram integrity is a significant problem for retailers.

Inventory Management
Inventory is a list of goods, materials or products, available in stock by a store (Saxena, 2009). Retail inventory management is the process and methods used to keep track of the stock in a retail store. These methods control everything from ordering, shipping, receiving, tracking inventory, retail turn-over, and storage. Retail inventory management can help keep a store's profits at a steady margin as well as reducing theft and loss of inventory. Many retail stores lose money every year because they lack a successful inventory management system in place. In later sections, we will discuss further on challenges associated with current retail inventory management system.

Warehouse Management
The warehouse management is a process of relevant warehousing activities such as product receipts, product issues, deliveries, internal and external stock transfers, product replenishment, warehouse inventory, supplier chain data synchronization, etc. The warehouse management is a huge and complex task, therefore mostly requires separate staff dedicated to run the warehouse.

Customer Service
The functions of customer service are very fragile, as they concern direct and indirect contact with store customers. Goals of customer service are to ensure meeting the needs of store customers with satisfaction and hence to maximize the sales. Retail customer service must:

- Provide efficient service to customers,
- Reduce waiting period at check-out counters,
- Reconcile transactions,
- Help customers find the desired product in the store and
- Furnish further information of the product, if asked.

Sales Management
The field of sales management involves the activities such as coordination of sales distribution, analyze and generate sales statistics, handling quotes and customer data decision-making for sales and discounts, etc.

The functional aspects of retail operations are highly interconnected and disturbance in one of the aspects can affect retail operations in whole and reduce store over-all earning. It is beyond the scope of our current work to address each aspect of retail operations. In this paper, we are focusing on problems of real-time planogram integrity, as related to inventory management, misplaced-item tracking and real-time information to the shopper and to retail operations.

3   The Role of a Planogram in Retail Operations

As consumers become active in more and more activities, they have less and less time available to shop in stores. Seizing the opportunity to create one-stop shopping environments to facilitate this experience, retailers are reacting by increasing store size to increase product assortments and categories. With larger product assortments, additional categories and more space for consumers to navigate, it is vital for retailers and vendors to direct and influence consumers' purchases. Thus, a planogram has become a vital tool to achieve consumer-centric store layout and product assortment in the shelf.

Retail stores keep track of their inventory through the concept of planograms. A planogram is a retailer's blueprint, which visually represents how the merchandise physically fit onto store shelf or fixture to allow proper visibility and price point options (Ray, 2010). A planogram displays how and where specific retail SKUs (stock-keeping unit) should be placed on retail shelves, racks, fixtures in order to increase consumer purchase. Fig. 2 shows different examples of planograms based on merchandise category.

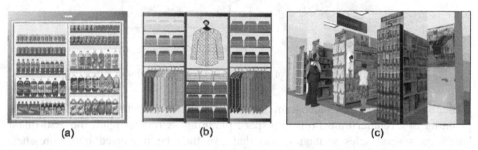

Fig. 2. Examples of planograms for (a) beverages (Source: Beverage, 2011), (b) clothes (Source: Cloth, 2012), and (c) a store (Source: 3D-planogram, 2012)

Planograms attempt to capture the absolute physical locations of an assortment (e.g., all hot beverages), the relative locations of items in an assortment (e.g., the relative locations of hot cocoa, tea and coffee), the amount of space allocated to each category (e.g., coffee), and each type of item (e.g., Starbucks coffee) within the category. Planograms can be store-wide (entire map of the store), aisle-specific (map of the refrigeration aisle), shelf-specific (map of the cosmetics shelf) or category-specific (map of where all coffee products are distributed in the store). There are multiple purposes for planograms:

- Retail operations: To help the store managers and store clerks know where items are located physically in the store, to keep track of inventory, to know where misplaced items should be located.
- Product placement and marketing decisions: To help the store decide where to place products, particularly new ones that the store is promoting. This helps the store decide how much to charge the vendor for product placement. The amount of available shelf space (particularly at key revenue-generation locations, such as

eye-level or at specific aisle end-caps) available is a marketing/revenue-based decision that the store must make on a regular basis.

- Customer experience: The appropriate layout of products can enhance "brand stickiness" and customer satisfaction. For example, ensuring that all associated product items (even if they are very different items and from different vendors) are within close proximity of each other is likely to reduce customer walk-aways rather than attempting to locate products by vendor or by type of item. For example, locating all baking-related items together, even if the vendors and items are diverse, is likely to improve the customer experience.

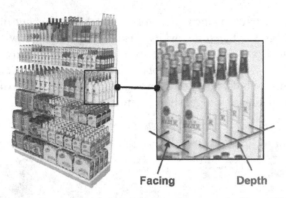

Fig. 3. Basic attributes of a planogram (Source: Attribute, 2012)

Basically, the design of planogram happens when the merchandise & assortment planning of a retailer transforms into space planning. A retailer plans the assortment based on several sales strategies, and that can then be translated into store-shelf planning as well. As shown in Fig. 3, there are two basic attributes of a planogram: facing and depth. Facing is the number of units of that particular SKU displayed in the front row of the shelf. Depth is the number of units of that particular SKU shelved one behind the other.

Apart from the basic attributes, there are other various important parameters, which are generally considered for SKU assortment and reflect the complexity of a planogram:

- Category of the SKU, e.g., Coke 300 ml can is affiliated with the Soft Drink Category, Cola Sub Category etc.
- SKU Details–SKU Name, UPC Code etc.
- Dimensions of the SKU - height, length, width, etc.
- Fixture details - type of fixture, location in the store, etc.

4   Challenges Associated with Retail Operations

Planograms are the result of much thought, research and effort to achieve a sales-and-profit goal. A retail study suggested that independent retailers can increase unit sales by 21%, sales by 12% and gross profit by 7% by maintaining planogram integrity (Food Distributors International, 2000).

Planograms get out-of-date due to several reasons: restocking, new products, seasonal changes in products and layouts, the last of these being fairly significant. For example, the layout of most supermarkets changes significantly in the month from Thanksgiving to Christmas, not just in terms of the items that are being displayed or promoted (e.g., cranberries during Thanksgiving vs. fruit-cake during Christmas), but also the layout of the store itself, particularly what is prominently displayed at the entry/access points of the store.

Stores construct planograms today in a time-consuming, manual and error-prone (not to mention, antiquated) manner. Incredibly, stores employ store clerks for manual walk-through and inventorying, on a shelf-by-shelf basis, at off-peak times in stores. This process, especially for large supermarkets, can be a significant, multi-week undertaking. The main problem is that a retail store planogram is often obsolete by the time it is fully constructed, lowering its utility, both for store operations as well as for revenue-generation and marketing.

Planogram Compliance
As one might imagine, implementing a store planogram is the toughest part of the retailing process. Retailers need to ensure that the new planogram arrived from the headquarters is being implemented in the store correctly. This verification process is called planogram compliance (Ray, 2010). Compliance with planograms is important, since optimal product placement can increase profit by up to 8.1% (Bishop, 2000). Planogram compliance is reported back to merchandising management and even nowadays to supply chain (Banker, 2011).

The research for ECR (efficient consumer response) in Europe found that 30% of the consumers would either substitute another brand or not buy at all, when their desired item is not found on the shelf. In 2002, Procter and Gamble (P&G)'s research showed that retailers lose 11% of sales due to out-of-stocks, and that same-brand substitutions win back less than 25% of those lost sales for a manufacturer (Procter and Gamble, 2002). Envisioning this problem, P&G embraced an approach "designing a supply chain from the shelf back", which generated $1B in 2004 (Smith, 2007). In this approach, supply chain works together with retailer to avoid the problems of out-of-stock items and temporary voids. In order to integrate such an approach, there are many challenges including cost of implementation for a technology in the shelf. There does not exist any reliable and cost-effective technology that can report planogram compliance to a supply chain in real-time.

Inventories
The average food retailer stocks 15000-60000 SKUs in thousands of categories and other retail channels are similarly glutted with items, a state of affairs exacerbated by approximately 10,000 new items introduced at retail every year (Industry Overview, 2010). The sheer volume of items strains the inventory-management systems for retailers and manufacturers alike, and drives up costs throughout the supply chain, including the retail, backroom, and warehouse levels (Byrnes, 2005).

Inventory-record inaccuracy, the discrepancy between the recorded inventory quantity and the actual inventory quantity physically available on the shelf, is a substantial problem in retailing. DeHoratius and Raman 2004 (Raman, 2004) found inaccuracies in 65% of the nearly 370,000 inventory records observed across thirty-seven retail stores. They identified several reasons for the 2 inaccuracies including replenishment errors, employee theft, customer shoplifting, improper handling of damaged merchandise, imperfect inventory audits, and incorrect recording of sales. A lack of inventory record accuracy clearly can reduce retail profits due to lost sales and labor and inventory carrying costs, which may run as high as 10 percent of existing profits (Raman, 2001).

In-Store Product Location Finding
According to the Wall Street Journal, the average user at a Wal-Mart Supercenter spends 21 minutes in the store, but locates only 7 of the 10 items on his or her shopping list (Wall Street Journal, 2007). Due to the same problem, stores like Home Depot and Best Buy are implementing measures to assist users in locating products, speed their checkout, and make their shopping experience less frustrating. Many consumers find this basic problem to be more frustrating in a food market where there are very few sales people roaming the aisles.

Product location can be even more frustrating in a grocery store where there are usually no sales people manning the aisles and very few stock attendants on the sales floor. Except for the addition of specialty departments such as bakeries, gourmet foods, and prepared foods, the basic grocery store design has not evolved much from its original configuration that assumed users would walk up and down every aisle on every visit because they would only shop once a week to stock up.

Currently, there is no legitimate solution for in-store product location finding. Shoppers generally need to rely on store staff to navigate them to the desired item. In case a shopper is unable to find a store staff, he/she might leave the store with disappointment and hold bad shopping experience. Eventually, the retail store looses the customer due to it inefficiency in providing production location service.

5   AndyVision: Transforming In-Store Retail Operations

Today, a planogram is a software platform that facilitates retailers to design store plan, fixture placements and SKU assortment planning. The complexity of the planogram software varies from store-to-store and shelf-to-shelf. There are numerous software packages in the market for planogram design including the 3D modeling of the store. The major problem is not associated with generation of the planogram, but with planogram integrity and compliance. First, a store receives the planogram from its headquarters or designs its own planogram, and then, the store staff have to regularly verify the product locations with their assigned locations in actual planogram. Planogram integrity or compliance check is the painstaking task of going from aisle-to-aisle and shelf-to-shelf, and assuring the product placement and its quantity. There is a close similarity between the methods of planogram compliance check and manual inventory count (often called "store sweeps"). In both methods, the store spends significant amount of time and money to ensure the quality of the retail operations.

There have been consistence efforts in the field of planogram technology; PlanoPad (Planorama Inc., USA) is a tablet-based planogram solution for planogram creation and compliance. ShelfSnap (ShelfSnap™, USA) is a camera based planogram compliance solution, which offers four-step approach; taking a photo of a shelf, transferring photos to the web, processing the data at server and providing data analysis. There are other number of similar solution providers for planogram design and compliance in the market. However, there is no existing technology, which by-passes the human for checking planogram integrity and automate the process.

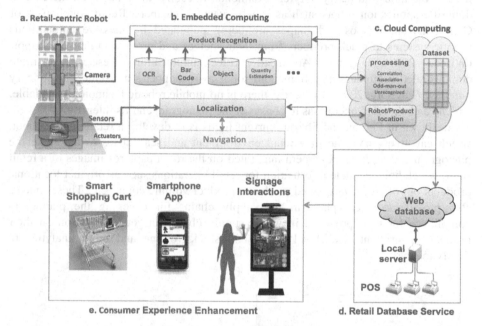

Fig. 4. AndyVision's automated planogram construction: (a) retail-centric robot, (b) embedded computing on the robot platform for product identification and robot localization, (c) cloud computing for accessing retail data base and advanced processing for product recognition and localization, (d) retail's existing database service, and (e) consumer experience enhancement through gesture-driven signage interaction and smartphone application.

AndyVision – An Approach for Automated Planogram Construction

The AndyVision project at Carnegie Mellon University aims to automate the process of retail planogram construction by deploying a retail-centric robot embedded with sensors, localization capabilities, and cameras. The robot scans the store aisles and generates real-time planograms for compliance check and inventory tracking. A cloud-computing platform provides the infrastructure for the big-data consumer-analytics processing as well as the data-intensive image stitching and image search/retrieval. In this approach, we also aim to enhance consumer experience via store digital signs and smartphone app. Shoppers can interact with the constructed planograms on the digital sign through intuitive hand and body gestures (and touch-based interactions, if needed) as well as browse the store products on a store mobile app. The architecture shows the unique capabilities of embedded-to-cloud interactions, and the power of enabling such interactions for transformative retail experiences.

Retail-Centric Robots

A recent report by the International Federation of Robotics found that 6.5 million robots serve humanity around the world (IFR, 2008). Large portion of the robots are used in the industrial heavy-duty environments. Recently, there has been increase in domestic application robots such as robotic vacuum cleaners. Retail warehouses of Gap, Amazon, Zappos and Stapes are being populated by massive autonomous robotic systems to pluck products from their shelves and provide in-house transport (Wired, 2009). Japan's Advanced Telecommunication Research Institute demonstrated usefulness of a humanoid robot as a shopper's assistant with grocery purchase (PhysOrg, 2009). Currently, there is no mobile robotic technology available, which can help retail operations to construct the planogram and its integrity.

As shown in Fig. 5a and 5b, our current focus is to develop a retail-centric robot, which can freely navigate in a retail environment and scan aisles to automate the planogram construction. A camera integrated on the robot captures images of a retail shelf, which hold valuable information for retail operations such as product locations, products out-of-stock, misplaced products, product volume count, etc. These images also provide excellent platform for supply chains for checking the planogram compliance for their products, in which supply chain can get information on their products assortment, shelf-level placement, stock volume and more analytics in timely manner.

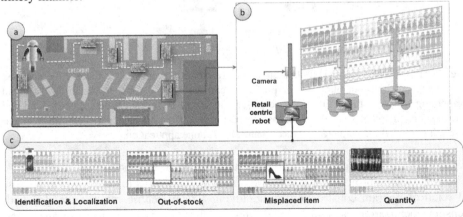

Fig. 5. AndyVision's robot (a) navigating inside a retail environment, and (b) scanning the products on a store shelf. Also shown are (c) use-cases requiring planogram integrity that AndyVision currently targets.

Embedded Computing

AndyVision's retail-centric robot is equipped with cameras for scanning retail environment, sensors for precise localization and actuators for in-store navigation. These sets of modules on the robot require on-board embedded computing to accomplish a huge task of planogram automation.

Product Identification

Previously, it has been noted that a typical food store stocks 15,000-60,000 SKUs on their shelves (Industry Overview, 2010). These SKUs vary from canned/bagged/boxed items to loose form. Identifying each and every item on the shelves by capturing an image is a challenging task and being under massive development. There are very few academic research groups, who pursue a task of in-store grocery identification, but with a different field of application. GroZi is an assistive technology for blind people to help them identify a product in the store and uses a computer vision approach for their technology (GroZi, 2007). Identifying a product in the store certainly needs integration of multiple techniques in computer vision:

- Optical Character Recognition (OCR): The OCR converts images with texts into machine-encoded text and can help identifying the product based on their name-tags on the product package. The name-tags or logos of the brands are normally unaltered and hence are a good source of information for categorizing a product.

Fig. 6. OCR for brand names

- Object Recognition based on attributes: Items such as loose vegetables/fruits, caps, purse, shoes etc. generally don't carry an identifiable logo or text and hence require different approach to achieve object detection with consideration of size, shape, color, texture, contour and other attributes. Fusion of various attributes has been shown to be powerful cues for object recognition (Bileschi 2005). Recently, Toshiba Tec., Japan has adopted a computer vision technology for fruit and vegetable detection at check out counter in order to avoid inputting the numbers (Toshiba 2012).

Fig. 7. SKUs for object recognition based on different attributes

- Template Matching:
  A method for template-based recognition of products/items includes capturing an image of the item packaging and extracting one or more attributes from the image. The attribute(s) is(are) used to find a matching template in a template image-set of grocery items. The matching template is then used to extract product information from the image. The web contains an immense collection of structured and unstructured image databases and offers the potential to generate useful models for image recognition and categorization (GroZi-120 Database, WebMarket Database).

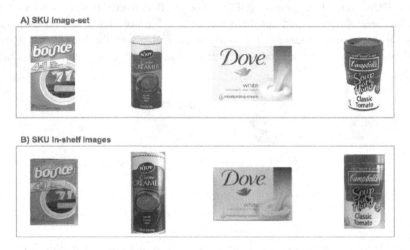

Fig. 8. SKU Template Matching: A) Image-set B) Test images

Robot Localization and Navigation

We have previously developed high-precision localization algorithms for tracking game-time assets (players, football, cameras, etc.) in the field of play for sports (Narasimhan, 2012). We plan to use this as a starting point for our in-store localization, particularly since our previous algorithms provided fine-grained localization. We describe our starting-point background work here, with respect to retail environments. Precise localization of the robot helps in generating a location-based inventory of the store items and in limiting the latency in product search on the database.

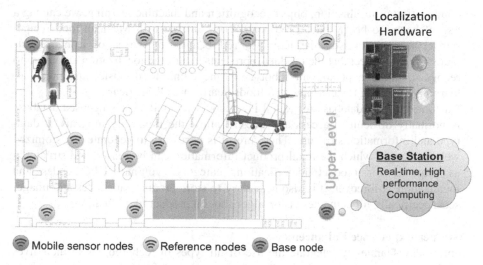

Mobile sensor nodes   Reference nodes   Base node

Fig. 9. In-store robot localization technology

There are three different types of nodes in this system based on their respective tasks - the Mobile Sensor Nodes (MSN) themselves are embedded within the planogram robot and also on restocking carts, the reference nodes are statically placed in the area of interest (the aisles, the aisle end-caps, and strategically located shelves) and a single base-node (a data aggregator, one for every collection of aisles/shelves, based on the specific store layout) collects the data from the reference nodes and transmits it to the cloud for further analysis. The base node gathers all the reference-node information during store operations, restocking times, etc., and estimates the location of the restocking carts and planogram robot, based on signal strength in real-time. The base station is an integral part of the back-end cloud platform.

Cloud Computing Back-End

Retail is a data-intensive environment and hence requires high performing connectivity to servers for data storage, image datasets, constructed planograms and other analytics. We categorize our cloud computing part into following subsections.

- Advanced Product Identification: Previously, we discussed embedded on-board product identification algorithms based on OCR, object recognition and template matching. Unfortunately, it is possible for these algorithms to have some errors, e.g., if on a shelf containing Campbell's cans of soup, some of the cans are turned around so that the labels cannot be read, or if some of the items cannot be identified because the product is placed in such a way that the label cannot be read. We will use cloud-based clustering algorithms to improve the hit-rate of product identification. Related products are often placed together, e.g., all of the Mexican foods are often placed together. If it is not possible to identify a specific product, we can perform clustering of nearby images to provide a coarse-grained classification, even if an exact one is not possible. For example, through a combination of localization, object-recognition and machine-learning, we can tag a bag of beans to be "Mexican red-beans" (or suitable to a Mexican-food recipe) rather than just classify the item as "red beans," by virtue of using object-recognition to detect that the product represents beans, the localization algorithm to recognize that the planogram robot is in the ethnic-foods aisle, and machine-learning/clustering to determine that foods nearby are all Mexican.

- Customized Web-database for Retail Products: The retail store's licensed database is normally difficult to access and doesn't allow the flexibility in-terms of data-type and information structure. Therefore, it is necessary to generate a customized web-service, in which the retail product information can be retried and structured as required for product identification, category, volume, UPC codes and localization. Moreover, it is also necessary to store products attributes and image-sets specifically for computer vision algorithm on product identifications.

Customer Experience Enhancements

People today seamlessly integrate the use of all types of technologies in their lives including the way they shop. As a result, they are more informed and selective about the products and services. In such an environment, the growth of mobile features and device convergence such as smartphones are driving mobile commerce. At the same time, store visits are enhanced via dynamic digital signs and personalization through hand-held devices or the shopper's own phone. These changes provide retailers with the opportunity to drive greater value by making the switch from "talking to" towards "engaging with" consumers and shoppers (Capgemini, 2012). Considering consumers as another channel apart from retail operations, we have stepped on number of technological developments for enhancing shopping experience.

Gesture-Driven Immersive Shopping on Digital Signs

Interactive digital signage at retail stores has increased value to the individual's shopping experience. Therefore, digital signage revenue will approach 4.5 billion in 2016 (ABI, 2011). To provide an interactive shopping experience, today's retail digital signs come with various technologies and sensing capabilities; face detection, gesture recognition, a large touch screen, etc. In our development with the digital sign, we are endeavoring to provide technology that facilitates shoppers not only to interact with the digital sign but also to navigate inside the store's 3D environment by hand gestures or using an App on their smartphone. The 3D environment of the store is generated from the planogram captured by the retail-centric robot.

Mobile Shopping Integration

Localizing an in-store item is very frustrating for a shopper. As our primary task is to generate planogram of the store shelves and automate the inventory count by finding a precise location of the items, we can use this information for assisting the shopper to locate an item in the store using their mobile phones. We are exploiting the accumulated data on the cloud and pinpoint the location of the items along with more augmented information.

6   Future Work: Pilot Deployment and User Studies

This technology will result in a living, working, fielded embedded system addressing a real pain-point for retail environments, with invaluable data from real users, developed and deployed with a willing and eager pilot partner, the Carnegie Mellon University Store. The Carnegie Mellon Store is a multi-purpose retail environment selling textbooks, trade books, merchandise, computer gadgets, university-branded clothing for men/women/kids, toys, trinkets, and student supplies.

Acknowledgments. This work was partially funded by the Intel Science and Technology Center in Embedded Computing (ISTC-EC). We are very thankful to our collaborators; Praveen Gopalakrishnan (Researcher, IntelLabs, Portland, USA), Ryan Wolfe (Director of Campus Services, Carnegie Mellon University, Pittsburgh, USA) and Angela Tumolo Neira (Marketing Specialist, University Store, Carnegie Mellon University, Pittsburgh, USA).

References

ABI. Digital Signage Market and Business Case Analysis, Research report, Q2 (2011)

Attributes. JDH Solutions, Category Management and Visual Management (2012), http://www.jdhsolutions.co.uk/Pages/default.aspx

Banker, S.: A robust merchandizing supply chain requires planogram compliance. Web-article (2011), http://www.logisticsviewpoints.com

Beverage, Planogram Gallery (2011), http://www.shelflogic.com/gallery.htm

Bileschi, S., Wolf, L.: A unified system for object detection, texture recognition, and context analysis based on the standard model feature set. In: British Machine Vision Conference (2005)

Bishop, W.: Documenting the value of merchandising. Technical report, National Association for Retail Merchandising Service (2000)

Capgemini, All-Channel Experience: Engaging with Technology Enabled Shoppers In-Store. Case Study (2012), http://www.capgemini.com (accessed on April 15, 2012)

Cloth, Retail Planograms (2012), http://www.dmsretail.com/retailplanograms.htm

Byrnes, J.: Achieving Supply Chain Productivity. Harvard Business Review Working Knowledge (2005), http://hbswk.hbs.edu/archive/4682.html

DeHoratius, N., Raman, A.: Inventory Record Inaccuracy: An Empirical Analysis. University of Chicago, Graduate School of Business (2004)

DeHoratius, N., Mersereau, A.J., Raman, A.: Retail Inventory management When Records are Inaccurate. University of Chicago, Graduate School of Business (2005)

Eagle, J.S., Joseph, E.A., Lempres, E.: From Products to Ecosystems: Retail 2010. McKinsey Quarterly (2000)

Fisher, M., Raman, A.: Introduction to focused issue: Retail operations management. Manufacturing & Service Operations Management. INFORMS 3(3), 189–190 (2001)

Zwiebach, E.: Shelf integrity seen as key to profit boost. Supermarket News (2000), http://supermarketnews.com/archive/shelf-integrity-seen-key-profit-boost

Gaffney, P.: Study Guide: Retail Merchandise Management, Chartered Institute of Purchasing and Supply. Chartered Institute of Marketing (2000)

Merler, M., Galleguillos, C., Belongie, S.: GroZi. Recognizing groceries in situ using in-vitro training data. In: International Workshop on Semantic Learning Applications in Multimedia (SLAM) (2007)

GroZi-120 Database, http://grozi.calit2.net/

Fisher, M.L., Raman, A., McClelland, A.S.: Rocket Science Retailing is Almost Here. Are You Ready? Harvard Business Review (July-August 2000)

Industry Overview, Supermarket Facts Industry Overview. Supermarket Facts, Food Marketing Institute (2010), http://www.fmi.org

Levy, M., Weitz, B.A.: Retailing Management. McGraw-Hill/Irwin, New York (1995)

Narasimhan, P.: Embedded Sports Technology: Taking it to the Field. In: Workshop on SoCs, Heterogeneous Architectures and Workloads, New Orleans, LO (2012)

Perdikaki, O.: Essays on Retail Operations. Ph.D. Dissertation, Kenan-Flagler Business School, University of North Carolina at Chapel Hill (2009)

Oxford Dictionaries, definition of planogram (2011)

PhysOrg, Supermarket Robot to Help Elderly (2009),
http://phys.org/news180261433.html

Procter and Gamble. Building a Smarter Supply Chain. Case study, Gartner Group (December 2002)

Raman, A., DeHoratius, N., Ton, Z.: Execution: the missing link in retail operations. California Management Review 43(3), 136–152 (2001)

Ray, R.: Supply Chain Management for Retailing. Tata McGraw Hill Pvt. Ltd. (2010)

Saxena, R.S.: Inventory Management controlling in a fluctuating demand environment. Global India Publications Pvt. Ltd (2009)

Smith, M., Poole, C.: Supply Chain Management. A report from Financial Times (2007)

Carr-Harris, D.: Toshiba. Supermarket checkout scans objects without barcodes (2012),
http://www.psfk.com/2012/03/
supermarket-scanner-no-barcodes.html

Wall Street Journal. Stores Help You Spend Money Fast. The Big Box Goal (June 2007)

WebMarket Database, http://yuhang.rsise.anu.edu.au/

Madrigal, A.: Wired. Autonomous Robots Invades Retail Warehouses. Online article on wired.com (2009)

3D-planogram, Free planogram software (2012),
http://www.visualsupercomputing.com/planogram-software.html

The Need for Standardized Tests
to Evaluate the Reliability of Data Transport
in Wireless Medical Systems

Helena Fernandez-Lopez[1,3], José A. Afonso[2,3], J.H. Correia[2,3],
and Ricardo Simoes[1,4,5]

[1] Institute for Polymers and Composites IPC/I3N, University of Minho, 4800 Guimarães,
Portugal
[2] Department of Industrial Electronics, University of Minho, 4800-058, Guimarães, Portugal
[3] Centre Algoritmi, University of Minho, 4800-058, Guimarães, Portugal
[4] Polytechnic Institute of Cávado and Ave, Campus do IPCA, 4750-810 Barcelos, Portugal
[5] Life and Health Sciences Research Institute (ICVS), School of Health Sciences,
University of Minho, Campus de Gualtar, 4710-057 Braga, Portugal
{hlopez,jose.afonso,higino.correia}@dei.uminho.pt
rsimoes@dep.uminho.pt, rsimoes@ipca.pt

Abstract. Wireless medical systems are comprised of four stages, namely the
medical device, the data transport, the data collection and the data evaluation
stages. Whereas the performance of the first stage is highly regulated, the others
are not. This paper concentrates on the data transport stage and argues that it is
necessary to establish standardized tests to be used by medical device
manufacturers to provide comparable results concerning the communication
performance of the wireless networks used to transport medical data. Besides, it
suggests test parameters and procedures to be used to produce comparable
communication performance results.

Keywords: eHealth, medical data transport, communication performance
assessment, standardized tests, wireless biomedical monitoring.

1   Introduction

For several decades, wearable patient monitors have been used in hospitals to
continuously monitor ambulatory patients. Electrocardiogram (ECG), oxygen
saturation (SpO_2), and other medical device information can be continuously
monitored, in real time, even when patients are on the move. Until recently, these
systems operated in dedicated spectrum bands and employed custom wireless
technologies designed to optimize specific quality of service (QoS) requirements[1].
However, advances in radiofrequency (RF) and networking technologies have

---

[1] QoS refers to the ability of a network to deliver data reliably and timely. According to the
IEEE Std. 11073-00101-2008, QoS requirements depend largely on the nature and criticality
of the data being transported, and include reliability, latency, priority, and bandwidth. Within
the scope of this work, the reliability of a data transport system refers to its ability to deliver
the generated packets as measured by, for example, the packet reception ratio.

F. Martins, L. Lopes, and H. Paulino (Eds.): S-Cube 2012, LNICST 102, pp. 137–145, 2012.
© Institute for Computer Sciences, Social Informatics and Telecommunications Engineering 2012

brought about new possibilities to healthcare providers and patients. For instance, pervasive networks based on the IEEE 802.11 protocols (commercialized as Wi-Fi) now approach the reliability of hardwired networks [1], allowing manufacturers to develop remote patient monitoring systems based on this protocol [2-4]. Besides, the rising costs in healthcare combined with significant developments in microelectronics, biomedical sensors and different classes of wireless networks decisively contributed to the increasing interest in e-Health systems[2] [5], including those that employ wireless technologies to transport medical data. These systems have the potential to reduce healthcare costs while improving the quality of the healthcare services provided to a huge number individuals, particularly elderly, recovering and chronically-ill patients [6].

However, despite the clear benefits of wireless data transport, several issues apply. The wireless channel is an unpredictable and challenging medium. First, the radio spectrum is scarce. Consequently, some frequency bands are shared by different systems resulting in interference between neighboring devices. Moreover, QoS parameters, such as data transfer reliability and latency, may also fluctuate in response to changes in traffic, link quality and propagation conditions. Finally, security is difficult to implement since wireless communications are susceptible to eavesdropping [7].

The incorporation of standard RF technologies into medical devices for wireless data transport has motivated standard organizations to address key aspects related to the transport of medical data. The CEN ISO/IEEE 11073-00101-2008 standard[3] [8] (hereinafter simply referred to as IEEE 11073-00101), which is part of the CEN ISO/IEEE 11073 family of standards, provides guidance for the utilization of point of care medical devices that exchange vital signs and other information using shared information technology infrastructure. This standard addresses technical and QoS performance requirements, besides security, privacy and coexistence issues. However, while requirements such as maximum end-to-end latency and bandwidth are clearly defined for various categories of medical data and scenarios, data transport reliability is only defined in qualitative terms, except for some equipment categories included in one specific scenario.

This limitation prevents the critical assessment of the suitability of emerging medical systems that employ wireless technologies to transport medical data. Besides, as there is no standardized evaluation test procedure that manufacturers should carry out, it is not possible to fairly compare the communication performance of different wireless medical systems. For instance, whereas some manufacturers assess the communication performance of a wireless network by measuring the total dropout proportion[4] [9], some academic works opt for determining the packet reception ratio (PRR) or its statistical distribution [10, 11]. Also, each study defines an arbitrary observation time.

This paper discusses the need of establishing standardized test procedures, to be used by medical device manufacturers, to provide comparable results about the communication performance of wireless networks used to transport medical data.

---

[2] According to the World Health Organization (WHO), eHealth is defined as the combined use of electronic communication and information technology in the health sector.

[3] CEN, ISO and IEEE stand for Comité Européen de Normalization or European Committee for Normalization, International Standards Organization, and Institute of Electrical and Electronics Engineers.

[4] The total dropout proportion equals to the total dropout time divided by the total monitored time.

Besides, it suggests test parameters and procedures that should be used to define standardized and repeatable tests that can produce comparable performance results.

2   Related Work

Sneha and Varshney [12] argue that the reliability of message delivery to healthcare professionals is the most critical requirement of patient monitoring. According to these authors, prioritization of different message types is vital to achieve high reliability. The reliability of message delivery on a wireless network can be evaluated through analytical models, simulations or experimental tests. In general, analytical models and simulations promise a fast evaluation that allows exploring the effect of relevant parameters and configurations. Although these performance evaluation approaches are steps towards obtaining insight into systems performance, an important further step is the execution of experimental tests [13]. However, if field tests are not standardized, it is difficult to compare the results obtained for different systems.

Differently from other works, this paper discusses the need for standardized test procedures for evaluating the reliability of wireless networks used to transport medical data. A similar effort in another area, the interoperability of information systems that exchange medical records, was recently carried out by the National Institute of Standards and Technology (NIST), in the USA, which elaborated several test procedures to improve the usability of Electronic Health Records [14].

3   Wireless Medical Data Transport Regulation

A wireless medical system can be described as being comprised of four stages [8], as shown in Fig. 1. The first stage includes the medical device, which can be an external device (e.g., a blood pressure monitor with wireless connectivity), a wearable or an implantable device. Data generated by devices are transported through multiple stages until reaching the patient or a health care provider.

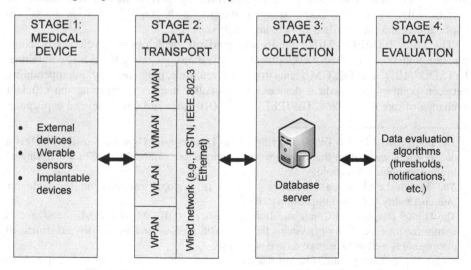

Fig. 1. Stages of a wireless medical system (adapted from [8])

Medical devices (stage 1 of Fig. 1) are regulated by federal government bodies. In most countries, these devices are classified into one of three or more classes. The regulatory requirements that manufacturers must comply with depend on the class which the device belongs. For instance, in the European Economic Area (EEA), medical devices can fall into one of the following classes: I, I sterile, I measure, IIa, IIb or III, with class III covering the highest risk products. Medical devices are regulated by three approach directives, depending on the classification of the device: Medical Device Directive 93/42/EEC [15]; Active Implantable Medical Device Directive 90/385/EEC [16]; or In-Vitro Diagnostic Medical Device Directive 98/79/EC [17]. These directives are in convergence to standards issued by the ISO, being the ISO 13485:2003, which defines the international quality system standards for medical devices, the most relevant [18]. These devices should also comply with product risk management, electromagnetic interference and compatibility (EMI/EMC) and usability regulations. Also, they must conform to local laws on personal data privacy and hardware disposal (Restriction of Hazardous Substances Directive or RoHS).

Whereas stage 1 of wireless medical systems is strictly regulated, the other stages are not. Current standardized technologies included in stage 2 were not designed to transport medical data and to support the QoS requirements that this transport involves. According to [8], such systems might not be considered a medical device when operating under typical conditions. Connectivity, including wireless connectivity based on standard-based technologies, is not considered on medical regulatory documents, but instead in nonmedical standards (e.g., IEEE standards) and nonmedical consortium agreements (e.g., Wi-Fi Alliance and ITU). The regulation within stage 2 would be hard to accomplish because data transport involves complex combinations of distinct technologies that include wireless personal area networks (WPANs), wireless local networks (WLANs), wireless metropolitan area networks (WMANs) and wireless wide area networks (WWANs). Data storage in stage 3 shares the same difficulties as stage 2, as it is defined by several nonmedical standards. Additionally, in several cases, stage 4 merges with stage 3 as it is based on applications that include data storage and analysis.

The CEN ISO/IEEE 11073 family of standards, which was developed in coordination with other standards development organizations, including IEEE 802 committee, IHTSDO[5], HL7[6] and DICOM[7], aims to provide real-time, plug-and-play[8] interoperability between point-of-care medical devices. Additionally, it aims to promote the efficient exchange of care device data. The IEEE 11073-00101 standard covers several application

---

[5] The International Health Terminology Standards Development Organization (IHTSDO) is a not-for-profit association that develops and promotes use of SNOMED CT, a multilingual health care clinical terminology.

[6] The Health Level Seven International (HL7) is a global authority on standards for interoperability of health information technology.

[7] The Digital Imaging and Communications in Medicine (DICOM) is a standard designed to ensure the interoperability of systems that deal with medical images and derived structured documents as well as to manage related workflow.

[8] Plug-and-play interoperability means that the user does not need to do any action, apart from connecting the device, to allow it to communicate data as defined.

use cases and considers potential applications for standard-based communication technologies, including cellular, IEEE 802.11, IEEE 802.15.4 and ZigBee protocols. Additionally, it defines QoS parameters, namely reliability, latency, priority, and bandwidth requirements, for some data classes of interest, as shown in Table 1. These parameters may be used in prioritizing and securing categories of data generated by devices that share a wireless network. As shown, real-time alarms and alerts should have the highest reliability and priority, whereas real-time patient state change messages and real-time reminders shall have highest reliability as well, but medium priority. Real-time waveform data and other real-time physiologic parameters, such as episodic blood pressure (BP) and heart rate (HR), shall have high reliability and priority.

Table 1. QoS requirements for some categories of medical device data (adapted from [8])

Data category	Reliability [*]	Latency	Priority [*]
Alerts/alarms (real-time)	highest/ essential	< 500 ms from the wireless sensor to the gateway to the wired network < 3 s communication latency	highest/ essential
Patient state change (real-time)	highest/ essential	< 3 s communication latency	medium
Reminder (real-time)	highest/ essential	< 3 s communication latency	medium
Waveforms (real-time)	high	< 3 s to central station < 7 s for telemetry to in-room monitor	high
Physiologic parameters (e.g., episodic BP, HR, SpO_2, temp.) (real-time)	high	< 10 s to central station < 3 s communication latency from monitor to clinician	high

(*) Ratings: low, medium, high and highest/essential.

As shown in Table 1, whereas maximum latency values are specified by the IEEE 11073-00101 for each category of data, reliability is only qualitatively specified (highest/essential, high, medium and low). The only exceptions are the maximum data losses allowed for some data categories included in the UC2 scenario, single cardiac patient in hospital, which are presented in Table 2.

Table 2. Data loss values for some data categories included in the UC2 scenario (adapted from [8])

Device	Data category	Data loss
PWD – Patient-worn device	1 ECG vector	< 5 seconds per event < 4 events per hour
Wireless vital signs monitor	PDS – Parameter data service (BP, HR, SpO_2, respiration)	< 5 seconds per event < 4 events per hour
	Alarm data services	< 1 second

Finally, the IEC 80001-1:2010 [19] defines roles, responsibilities and activities that are necessary for risk management of IT-networks incorporating medical devices to address safety, effectiveness and data and system security. However, it applies after the medical device has been acquired by the organization and does not define minimum performance parameters.

4   Assessment of the Communication Performance of Networks Used to Transport Medical Data

In a recent work, the authors have evaluated the communication performance of a prototype remote patient monitoring system consisting of six ZigBee-based ECG devices [20]. Fig. 2 presents the per 2-second PRR values measured for a specific wireless ECG sensor device over a period of 16.7 hours. This device was two hops away from the personal area network (PAN) coordinator and achieved a mean PRR of 0.99. However, as shown in Fig. 2, during a contention period of approximately 30 minutes, the PRR values varied considerably, reaching a minimum of 0.6, which, despite the good mean PRR, might be unacceptable for certain scenarios.

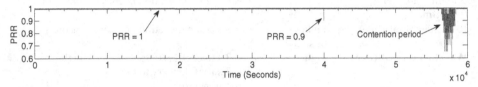

Fig. 2. PPR values for a ZigBee-based ECG device measured over a period of 16.7 hours

As illustrated, test results expressed using mean values alone (e.g., mean PRR, mean dropout duration, mean time between dropouts and total dropout proportion) are unable to provide complete information about the reliability of a wireless network. However, mean values are often presented by several studies, including those provided by medical device manufacturers. The occurrence of transitory low communication performance periods, as the one exemplified on Fig. 2, may not become evident to a clinical engineer in charge of evaluating the performance of a wireless medical device if PRR values along the time are not provided. Besides, if evaluation methods are not standardized, it may not be possible to fairly compare the communication performance of different systems.

Consequently, it is necessary to define standardized tests that can provide, to all involved stakeholders, essential information about the communication performance of the wireless networks used to transport medical data. In a first approach toward solving the identified problem, it is suggested that performance assessment tests include the general aspects presented in Table 3. They contain different wireless channel conditions, mobility scenarios, scalability issues, and failure conditions. Also, it is recommended using a test platform that guarantees controlled conditions in order to assure repeatability. Such a platform was presented by Hu et al., from National ICT Australia and University of Queensland [21], who evaluated the performance of specific IEEE 802.11a-based mesh backhaul radios. The referred platform provided an environment free of interference. Besides, it allowed researchers control the quality of all links and simulate link and device failures.

Table 3. General procedures suggested for performance assessment of wireless networks used to transport medical data

Aspects	Procedure
Wireless channel conditions	Ideal propagation: consider free-space path loss determined using the maximum range specified for the patient-worn device.
	Non-ideal propagation conditions: consider additional attenuation caused by different effects (e.g. multipath, interference and obstruction caused by large objects and structures).
Mobility	Consider one or more mobile patient-worn devices which move from one parent device/sink to another one.
Scalability	Add patient worn devices until the maximum number of devices supported. Consider situations where devices should associate to the same or to different parent devices/sinks. Also, for multihop networks, consider a crescent number of hops.
Backhaul device failure, link failure and connection loss	Measure the time necessary to reassociate to a new parent device/sink in case of backhaul device or link failure. Consider reliability requirements of different data categories and distinct reconfiguration actions.
	In case the device moves to an area without connectivity, verify if actions taken are acceptable (e.g., alerts to care givers and patient).

In order to correctly evaluate the communication performance of a medical device, it is suggested keeping track of the following QoS parameters:

- PRR measured over time using a temporal window of appropriate length for each active patient worn-device and medical data category (refer to Table 2 of [8] for the comprehensive list of medical data categories);
- overall and per-active patient-worn device average dropout;
- end-to-end latency associated to each message and the mean and maximum end-to-end latency values for each medical data category;
- jitter incurred by periodic traffic with constant packet size; and
- bandwidth required per active patient-worn device.

As stated above, it is suggested measuring the PRR over a small time window in order to capture transitory effects that compromise the system performance. For instance, in carrier sense multiple access with collision avoidance (CSMA-CA) based networks transitory contention conditions between devices may result in temporary message losses generated by contending devices. Besides, reconfigurations, such as rate changes from multi-rate mechanisms used to adapt the transmission rate to the channel condition in IEEE 802.11-based networks, can temporarily impact the performance of the active devices involved.

During the lifetime of the equipment, it is essential that field tests are performed. Ideally, the medical system would incorporate a test module capable of acquiring data that would alert clinical staff of any changes in the reliability of the system. Also, it would provide clinical engineers means to execute field tests and, in case of any trouble, supply the detailed information needed to diagnose the problem. In case of routinely field tests required for system evaluation, these tests should last long enough to capture most variations in environmental conditions. For instance, in a hospital or nursing home scenario, a field test may last one week to capture events that can influence the RF environment. These events would include the increased influx of visitors during the weekends which, in general, have negative effects on link conditions.

5   Conclusions

This paper discusses the lack of quantitative reliability parameters to evaluate the communication performance of medical systems that employ wireless communication technologies. The IEEE 11073-00101, which provides general guidance for the use of standard RF technologies to transport medical data, establishes maximum latency values for each data type. A similar approach is needed for reliability.

Moreover, it demonstrates the need for standardized and repeatable tests to be used by medical device manufacturers to produce comparable communication performance evaluation results regarding medical systems that use wireless networks to transport data. In order to provide an initial contribution to this field, it suggests general test parameters and procedures. The QoS parameters to be measured include the instantaneous and mean PRR per-active patient-worn device and the bandwidth required by each device. The general procedures include measurements performed under different channel conditions and mobility scenarios. Also, it suggests addressing scalability and different failure conditions, such as backhaul device failure, link failure and connection loss.

Future work includes the communication performance assessment of a prototype wireless medical system using the guidelines proposed in order to evaluate the suggested approach.

Acknowledgments. Authors would like to thank Mr. Teófilo Leite, Mr. Nélson Brito and Ms. Teresa Moura, from Hospital Privado de Guimarães, for the encouragement and support. This work has been supported by the Portuguese Foundation for Science and Technology (FCT), Lisbon, through the 3° Quadro Comunitário de Apoio, the POCTI and FEDER programs, the MIT-Portugal program, project PEst-C/CTM/LA0025/2011 (Strategic Project - LA 25 - 2011-2012), and the FCT grant SFRH/BD/39408/2007. AAL4ALL project, co-financed by the European Community Fund through COMPETE - Programa Operacional Factores de Competitividade. Clinical and financial support for this study has been provided by Grupo AMI – Assistência Médica Integral (Casa de Saúde Guimarães, SA), Portugal, under the partnership established between this healthcare company and the University of Minho. We also acknowledge useful discussions with staff from the Hospital de Braga.

References

1. Baker, S.D., Hoglund, D.H.: Medical-Grade, Mission-Critical Wireless Networks (Designing an Enterprise Mobility Solution in the Healthcare Environment). IEEE Engineering in Medicine and Biology Magazine 27, 86–95 (2008)
2. Allyn, W.: Acuity Central Monitoring Station, http://www.welchallyn.com/wafor/hospitals/emr_connectivity/solutions_WACS.htm
3. Philips Healthcare. Philips Intellivue Telemetry System x40, http://www.healthcare.philips.com/main/products/patient_monitoring/products/intellivue_mx40/index.wpd
4. Intelesens. Aingeal Patched-based Vital Signs, http://www.intelesens.com/pdf/aingealdatasheet.pdf
5. Kyriacou, E., et al.: m-Health e-Emergency Systems: Current Status and Future Directions. IEEE Antennas and Propagation Magazine 49, 216–231 (2007)
6. Ekeland, A.G., Bowes, A., Flottorp, S.: Efectiveness of telemedicine: A systematic review of reviews. International Journal of Medical Informatics 79, 736–771 (2010)
7. Goldsmith, A.: Wireless Communications. Cambridge University Press, New York (2005)
8. IEEE 11073-00101-2008 Health Informatics - PoC Medical Device Communication - Part 00101: Guide - Guidelines for the Use of RF Wireless Technology, 1–99 (2008)
9. Allyn, W.: FlexNet for 802.11a life-critical wireless networks, http://www.welchallyn.com/products/en-us/x-16-vo-96-1234190869014.htm
10. Ko, J., Gao, T., Terzis, A.: Empirical Study of a Medical Sensor Application in an Urban Emergency Department. In: Proceedings of the Fourth International Conference on Body Area Networks, ICST, pp. 1–8. ICST, Belgium (2009)
11. Chipara, O., Lu, C., Bailey, T.C., Roman, G.-C.: Reliable Patient Monitoring: A Clinical Study in a Step-down Medical Unit. Technical report, Washington University in Saint Louis, Saint Louis (2009)
12. Sneha, S., Varshney, U.: Enabling ubiquitous patient monitoring: Model, decision protocols, opportunities and challenges. Decision Support Systems 46, 606–619 (2009)
13. Ali, M., et al.: Medium Access Control Issues in Sensor Networks. ACM Computer Communication Review 36, 33–36 (2006)
14. National Institute of Standards and Technology (NIST), http://www.nist.gov/
15. Council Directive 93/42/EEC of 14 June 1993 concerning medical devices (1993)
16. Council Directive 90/385/EEC of 20 June 1990 on the approximation of the laws of the Member States relating to active implantable medical devices (1990)
17. Directive 98/79/EC of the European Parliament and of the Council of 27 October 1998 on in vitro diagnostic medical devices (1998)
18. ISO 13485:2003 Medical devices - Quality management systems - Requirements for regulatory purposes, 1– 57 (2003)
19. IEC 80001-1:2010 Application of risk management for IT-networks incorporating medical devices - Part 1: Roles, responsibilities and activities, 1–70 (2010)
20. Fernandez-Lopez, H., Afonso, J.A., Correia, J.H., Simoes, R.: HM4All: A vital Signs Monitoring System based in Spatially Distributed ZigBee Networks. In: 4th International Conference on Pervasive Computing Technologies for Healthcare, pp. 1–4 (2010)
21. Hu, P., et al.: Evaluation of commercial wireless mesh technologies in a public safety context: Methodology, analysis and experience. In: IEEE 7th International Conference on Mobile Ad Hoc and Sensor Systems, pp. 606–611 (2010)

Link Stability in a Wireless Sensor Network –
An Experimental Study

Stefan Lohs, Reinhardt Karnapke, and Jörg Nolte

Distributed Systems/ Operating Systems group,
Brandenburg University of Technology,
Cottbus, Germany
{Slohs,Karnapke,Jon}@informatik.tu-cottbus.de

Abstract. Most routing protocols proposed for wireless sensor networks
are based on the standard approach also used in many other types of
networks, e.g. MANETS, even though the conditions are drastically different. To evaluate the usefulness of reactive routing protocols based on
route discovery by flooding of route request messages it is necessary to
understand the nature of the underlying wireless communication links.

In this work we present the results of connectivity measurements conducted with current sensor node hardware, taking special interest in the
number of unidirectional links present and the frequency of link changes.

Keywords: wireless sensor networks, link stability, unidirectional links.

1 Introduction

The most common way to search for a route in a wireless network is to flood a
route request message from the source to the destination, which then answers
with a route reply message. In some protocols, this route reply is transmitted
using the inverted path taken by the fastest route request (e.g. AODV [7]),
in others, it is also flooded and may take a different route entirely (e.g. DSR
[5] when using unidirectional links). Sometimes a second route reply from the
originator of the route request is necessary. In some other protocols other names
may be used, but the basic mechanism is the same.

All of these protocols are based on one assumption: If a route can be discovered
by these flooding mechanisms at some time, the route can be used at least for a
certain time. If routes break due to link changes, different handling methods are
used. But these route maintenance mechanisms are often expensive, and should
be performed as seldom as possible. Therefore, a certain link stability is the basic
requirement for these protocols to perform according to their specification.

Another frequently made assumption is that unidirectional links are uncom-
mon and that it is better to ignore them when making routing decisions [6].

In this paper we present connectivity measurements conducted with real sensor network hardware and show that unidirectional links are not only common,
but their number exceeds that of bidirectional links by far. Also, we show that

F. Martins, L. Lopes, and H. Paulino (Eds.): S-Cube 2012, LNICST 102, pp. 146–161, 2012.
© Institute for Computer Sciences, Social Informatics and Telecommunications Engineering 2012

the number and frequency of link changes is even higher than expected, making changes to the way routing information is handled in traditional routing protocols necessary.

This paper is structured as follows: Related work, which gave us the idea that it would be worthwhile to investigate connectivity between nodes, is shown in section 2. Section 3 describes the sensor nodes from Texas Instruments we used for our experiments. The gathered connectivity information is shown in section 4. We finish with a conclusion in section 5.

2   Related Work

2.1   The Heathland Experiment: Results and Experiences

The authors of [10] describe an experiment they conducted in the Lüneburger Heide in Germany. The original goal was to evaluate a routing protocol, which is not characterized further in the paper. Rather, the observations they made concerning the properties of the wireless medium are described, focusing on the frequency of changes and the poor stability of links. These experiments were conducted using up to 24 Scatterweb ESB [9] sensor nodes, which were affixed mostly to trees or poles, and left alone for two weeks after program start. One of the duties of the network was the documentation of the logical topology (radio neighborhood of nodes), which was evaluated by building a new routing tree every hour, e.g. for use in a sense-and-send application. The neighborhood was evaluated using the Wireless Neighborhood Exploration protocol (WNX) [10], which can detect unidirectional and bidirectional links. All unidirectional links were discarded and only the bidirectional ones used to build the routing tree.

Figure 1(a) shows one complete communication graph obtained by WNX, while figure 1(b) shows the same graph without unidirectional links. Here, a lot of redundant paths have been lost by the elimination. In fact, one quarter of the nodes are connected to the rest of the network by merely one link when

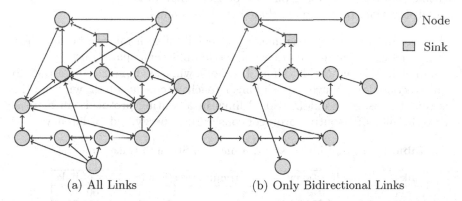

(a) All Links   (b) Only Bidirectional Links

Fig. 1. A Communication Graph (a) with and (b) without Unidirectional Links (taken from [10] presentation: [11])

unidirectional links are removed. If this single link breaks, the nodes become separated, even though there are still routes to and from them. Thus, the removal of unidirectional links increases the probability of network separation.

2.2 On Exploiting Asymmetric Wireless Links via One-Way Estimation

The authors of [8] propagate a similar opinion. They evaluate three kinds of links (asymmetric, unidirectional, bidirectional) using protocols like ETX (Expected Transmission Count) [2]. These protocols search for reliable connections, but most of them only focus on bidirectional links. This leads to the fact that a link with a reliability of 50% in both direction is chosen above one with 100% from node A to node B and 0% from B to A. If data needs to be transmitted only from A to B without need for acknowledgment, this choice is obviously wrong. To prevent this wrong choice, the authors of [8] propose a protocol called ETF (Expected Number of Transmissions over Forward Links), which is able to use unidirectional links. They also show that the reach of reliable unidirectional links is greater than that of reliable bidirectional links.

In experiments with XSM motes [8] 7 × 7 nodes were placed in a square, with a distance of about 1 meter between nodes. In four sets of experiments at different times of day each node sent 100 messages at three different power levels. Then the packet reception rate was recorded. It is defined for a node A as the number of packets A received from a node B divided by the number of messages sent (100). Afterwards, the packet reception rates of nodes A and B are compared. If the difference is less than 10%, the link is considered bidirectional. If it is more than 90% the link is considered unidirectional. The XSM nodes offer nine different transmission strengths, of which three were evaluated: the lowest, the highest and the third in between. Table 1 shows the results of the experiments.

The results show that even when using the maximum transmission strength, 12% of the links would have been discarded by ETX and similar link quality evaluation protocols that focus only on bidirectional links.

The observations of [8] are concluded in three points:

1. Wireless links are often asymmetric, especially if transmission power is low
2. Dense networks produce more asymmetric links than sparse ones
3. Symmetric links only bridge short distances, while asymmetric and especially unidirectional ones have a much longer reach. A conclusion drawn from this is that the usage of unidirectional links in a routing protocol can increase the efficiency of a routing protocol considering energy and/or latency.

Table 1. Link Quality versus Transmission Strength (taken from [8])

link quality	bidirectional	asymmetric	unidirectional	number of links
power level 1	50%	43%	7%	500
power level 3	65%	22%	13%	1038
power level 9	88%	6%	6%	1135

2.3 Design and Deployment of a Remote Robust Sensor Network: Experiences from an Outdoor Water Quality Monitoring Network

A sensor network which monitors water pumps within wells is described in [3]. The sensors were used to monitor the water level, the amount of water taken and the saltiness of the water in a number of wells which were widely distributed. The necessity for this sensor network arose because the pumps were close to shore and a rise in saltiness was endangering the quality of the water. The average distance between wells was 850 meters and the range of transmission was about 1500 meters. Communication was realized using 802.11 WLAN hardware both for the nodes as well as for the gateway. For data transmission between nodes *Surge_Reliable* [13] was used, which makes routing decisions based on the link quality between nodes.

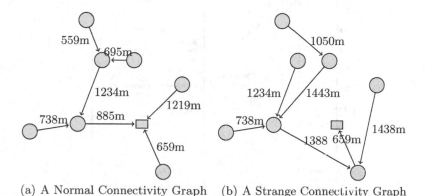

    (a) A Normal Connectivity Graph   (b) A Strange Connectivity Graph

Fig. 2. Two Communication Graphs (taken from [3])

During the experiments, it could be seen; that the (logical) topology of the network changed dynamically, even though all nodes were stationary. The authors claim that these changes were probably due to antenna size and changes in temperature and air moisture. In this context it is important to remember that the distance of nodes was far below the range of the transmitters (about 50%). While about 70% of the routing trees observed followed the theory (figure 2(a)), there were a lot of strange topologies. In one case the average distance between connected nodes even rose to 1135 meters, as nodes that should have been able to communicate directly with the gateway were connected to nodes on the far side instead. In one of these routing trees (figure 2(b)), a single node had to take care of all communication with the gateway, even nodes that were on the other side were using it as next hop.

2.4   Taming the Underlying Challenges of Reliable Multihop Routing in Sensor Networks

The main focus of [13] is link quality estimation. The authors measured link quality for a sensor network deployment consisting of 50 MICA motes from Berkeley.

Figure 3 shows the results they obtained. All nodes within a distance of about 10 feet (about 3 meters) or less from the sender received more than 90% of the transmitted packets. The region within 10 feet of the sending node is therefore called the effective region. It is followed by the transitional region. Nodes in this region can not be uniformly characterized as some of them have a high reception rate while others received no packets at all. In the transitional region, asymmetric links are common. The last region is the clear region and contains only nodes that did not receive any transmissions.

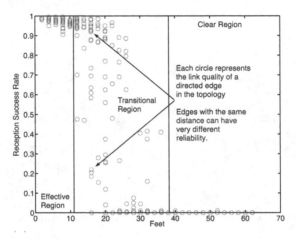

Fig. 3. Effective, Transitional and Clear Region (taken from [13])

2.5   Understanding Packet Delivery Performance in Dense Wireless Sensor Networks

The authors of [14] measured the properties of wireless sensor networks on the physical and medium access control layers. These measurements were conducted using up to 60 MICA motes, which were placed in three different environments: An office building, a parking lot and a habitat. The experiments for the physical layer were realized with a single sender and multiple receiver nodes and have shown the existence of a *grey area* in reception which can consist of up to one third of the network. In this grey area, the reception quality of nodes varies a lot, both spatial as well as in time. This observation is similar to the *transitional region* described in [13](see section 2.4). Another result described by the authors is that in the parking lot and indoor environments nearly 10% of links are asymmetric. Please note that what the authors call asymmetric links is otherwise referred to as unidirectional links in this paper: *"Asymmetry occurs when a node can transmit to another node but not vice versa"* [14].

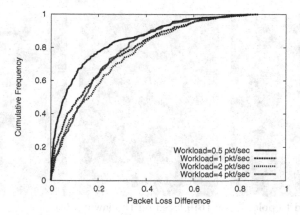

Fig. 4. Packet Loss Difference for Pairs of Nodes (taken from [14])

Figure 4 shows the results obtained by the authors of [14]. They have defined the packet loss difference for two nodes as the difference between the packet delivery efficiency of both nodes. The figure shows that asymmetric (unidirectional) links are quite common: More than 10% of the surveyed links have a difference of more than 50%.

The final claim the authors make is about asymmetric (i.e. unidirectional) links: *"The fraction of asymmetric links is high enough that topology control mechanisms should, we argue, carefully target such links"*.

3 Hardware and Application

The eZ430-Chronos from Texas Instruments [4] is an inexpensive evaluation platform for the CC430. It features an MSP430 microcontroller with integrated CC1100 sub-gigahertz (868MHz) communication module [1]. The evaluation board is delivered as a compact sports watch containing several sensors, e.g. a three-axis accelerometer and 5 buttons which are connected through general purpose I/O pins. The sports watch casing has been removed in order to use the eZ430s as sensor nodes.

Figure 5 shows the used eZ430-Chronos sensor nodes in three different placements (see below). An external battery pack has been soldered to the nodes, which replaces the internal coin cells. This enables the usage of freshly charged batteries for each experiment.

To get a feeling for the behavior of the real hardware and to keep the possibilities of application errors to a minimum, the first experiments were made using a fairly simple application. 36 sensor nodes were deployed on the lawn outside of the university's main building, spaced one meter from each other.

As only the connectivity should be measured, the "application" consisted only of a flooding with duplicate suppression. Node 0 was connected to a laptop via USB and transmitted 50 messages containing a sequence number (increased every round) and the ID of the last hop, with a one minute pause between messages.

(a) affixed to poles   (b) placed on the lawn   (c) on a stone pavement

Fig. 5. A modified eZ430-Chronos Sensor Node

Each node that received a flooded message first logged the neighbor from which it received the message. After that the node checked if it had already handled a message with this sequence number. If it had, the message was discarded, otherwise the node changed the field *last hop* to contain its own ID and rebroadcast the message. This change of the last hop in the message enables the detection of all incoming links by each node. The decision whether a link was unidirectional or bidirectional was made offline, once all connectivity data had been gathered and combined. If node A had an incoming link from node B for sequence number X and node B had an incoming link from node A for that same sequence number X, the link is considered bidirectional at time X.

Even this simple application ran into two problems: The CC430 uses a so-called CCA Medium Access Control, which is basically a CSMA/CA scheme. A node that wants to transmit a message waits for a random time (backoff) before sensing the medium. If it is free, the message is transmitted. Otherwise, the radio waits for a random time before trying again. The used hardware was not able to receive messages during the backoff, which meant that even in an experiment with 3 nodes (0, 1, 2) node 2 was never able to receive messages from node 1, because it was still in its backoff when node 1 transmitted. To solve this problem for the connectivity evaluation, a software delay was introduced. The software waited between 1 and 13 milliseconds before passing the message to the hardware. This delay could be tolerated, because application knowledge was available (node 0 transmitted a new packet only every minute).

Retrieval of data was induced by sending a message to a node, telling this node that it should transmit its gathered neighborhood information. But the nodes were not able to receive any messages after a seemingly random time. Sometimes, nodes functioned only for a couple of minutes, sometimes nodes ran for more than a day and still responded. The influence of stray messages on the application can be ruled out due to precautions in the software. The problem seems to exist in the state machine of the radio. To remove this problem, a watchdog timer was introduced which resets the radio every 5 minutes if the application did not receive any messages during that time. If it did receive a message, the watchdog was restarted. While this could lead to problems if the nodes radios failed during the experiment, it was mainly used to gather the results, once the sensor nodes were collected and returned to the office.

4   Results

Four different placements were evaluated: On a lawn, on a stone pavement, affixed to poles and taped to trees. The first three placements were also evaluated on two different radio channels.

4.1   Lawn Experiment, Channel 0, Sink (Node 0) Connected to Laptop

The first experiment was conducted on the lawn in front of our university.

Figure 6 shows the connectivity graph obtained for the first of the 50 messages that were flooded into the network. One of the nodes, node 30, had a defective contact and did not participate at all. Four other nodes, nodes 12, 27, 28 and 33 suffered a complete reset during transportation, leading to loss of the connectivity data they gathered. Still, this had no effect on the network load at runtime, and a lot of information could be gathered.

Node 0, which was connected to a laptop using an USB cable, was heard by lots of nodes, even those far away like node 11, node 29 or node 31. This shows that the transmission strength of the nodes, while it was set to 0 dBm for all nodes, still depends on the power supply of the nodes, i.e. the batteries. In deployments where a sink node connected to a fixed power supply such as a computer should be used, the longer reach of the sink node might well be a problem. This problem

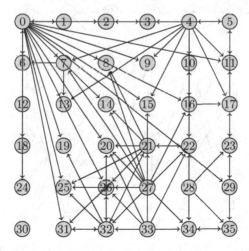

Fig. 6. First Connectivity Graph obtained on the lawn, channel 0

would for example manifest, when a tree routing approach is used, and the sink floods a message through the network to establish initial parent and child nodes, as most of the nodes would assume node 0 as their parent, but be unable to transmit directly to it.

Also, the results show that even though the nodes were only one meter apart from each other, bidirectional links are rare and unidirectional links are common. If all links are counted, 3018 unidirectional and only 403 bidirectional links have been recorded during the 50 minute deployment. If the unidirectional links from the nodes that have failed during transport are deducted (560 seemingly unidirectional ones), the ratio is still 2458 unidirectional links against 403 bidirectional ones. 7019 link changes occurred.

To remove the influence of the higher transmission strength of the "sink" (node 0), all links to and from node 0 can be removed from the equation. But even then, the result seems pretty obvious: 1477 unidirectional links stand opposed to 355 bidirectional ones (ratio 4.16 : 1).

4.2 Lawn Experiment, Channel 0, Sink (Node 0) Connected to Batteries

To remove the influence of the USB cable connected to node 0, the experiment was repeated. This time, and in all subsequent experiments, node 0 used a normal battery pack like all the other nodes. Even though precautions were taken, one node (node 25) still suffered a reset before the gathered data could be retrieved. The application was the same, with 50 flooded messages. 4039 unidirectional links as well as 818 bidirectional links were recorded this time, if the links from node 25 are removed that still leaves 3912 unidirectional ones opposing 818 bidirectional links (4.78 : 1 ratio) with 7019 link changes over the length of the whole experiment. The first connectivity graph is presented in figure 7.

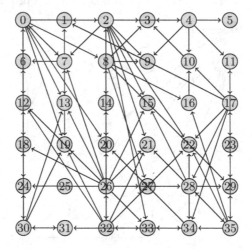

Fig. 7. First connectivity graph obtained on the lawn with node 0 connected to batteries, channel 0

4.3   Stone Pavement Experiment, Channel 0

To evaluate the influence of the ground on which the sensor nodes were placed, the experiments were repeated again, but this time the nodes were placed on the stone paved yard of the university. Figure 8 once again shows the first connectivity graph obtained.

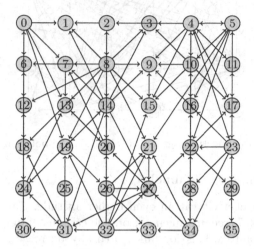

Fig. 8. First Connectivity Graph obtained on the stones on channel 0

Altogether 3570 unidirectional links and 851 bidirectional ones were measured, resulting in a ratio of 4.19 : 1, which is nearly the same as the ratio obtained in the first row of experiments and only a little bit different from the results of the second row of experiments. The average ratio seems to be between 4 and 5 to 1, even though individual values vary between 2.40 to 1 and 11 to 1. 6589 link changes occurred.

4.4   Pole Experiment, Channel 0

The previous three rows of experiments were all conducted with sensor nodes that lay on the ground, which is a safe assumption for many deployments. However, if the nature of radio communication is taken into account, the nodes should be placed with a certain distance from the ground; to increase the communication range and reception. Therefore, the 36 sensor nodes were connected to wooden poles and placed about 20 cm above the university lawn in these experiments.

Figure 9 visualizes the first obtained connectivity graph.

Altogether 5150 unidirectional links and 492 bidirectional ones (ratio 10.47 : 1) with a total of 7146 changes were measured. Interestingly, the better radio characteristics increased the number of unidirectional links far more than the number of bidirectional ones. The ratio of unidirectional ones to bidirectional ones increased up to 18 : 1.

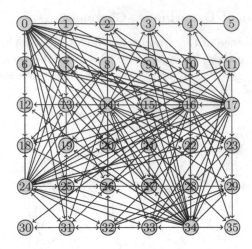

Fig. 9. First connectivity graph obtained on the poles on channel 0

4.5   Tree Experiment, Channel 0

To evaluate the connectivity at an even higher altitude, the sensor nodes were next fitted to a 5 × 5 tree arrangement on the campus of our university. Please note that the absolute values for links naturally decreases, as only 25 nodes are used in this scenario, instead of 36. Figure 10 shows the initially measured connectivity.

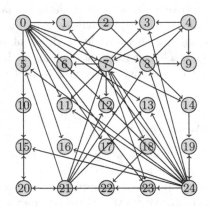

Fig. 10. First Connectivity Graph obtained on the trees on channel 0

A total of 2977 unidirectional links and 330 bidirectional ones were measured (ratio 9.02 : 1) with 3329 link changes occurring during the 50 minutes runtime of the experiment.

4.6   Lawn Experiment, Channel 3

To evaluate the influence of the chosen channel on the connectivity, the experiments on the poles, the lawn and the stone pavement were repeated on channel 3.

The initial connectivity graph is shown in figure 11.

As much as 4411 unidirectional links and 757 bidirectional ones (ratio 5.83 : 1) with 7103 link changes were measured.

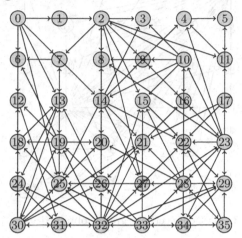

Fig. 11. Initial Connectivity on the Lawn, September 23rd, 2011

4.7   Stone Experiment, Channel 3

The initial connectivity graph obtained on the stone pavement is shown in figure 12. A total of 3508 unidirectional links and 712 bidirectional ones were detected (ratio 4.93 : 1), with 5528 link changes occurring.

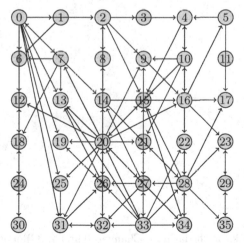

Fig. 12. Initial connectivity graph on the stone pavement

4.8   Pole Experiments, Channel 3

The first connectivity graph obtained is visualized in figure 13.

Altogether 4761 unidirectional links and 225 bidirectional ones (ratio 21.61 : 1) with 5541 changes were measured.

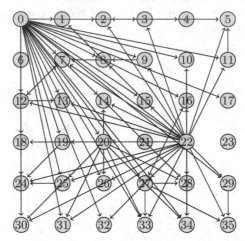

Fig. 13. First connectivity graph, obtained using poles

4.9   Summary

The connectivity measurements have shown that unidirectional links occur even more often than literature suggests and that the height of the placement does influence the communication range. More specific, the number of unidirectional links increases stronger than the number of bidirectional ones.

Figure 14 shows the results of the lawn experiments on channels 0 and 3 in detail. Each round represents one flooded message, with one minute passing

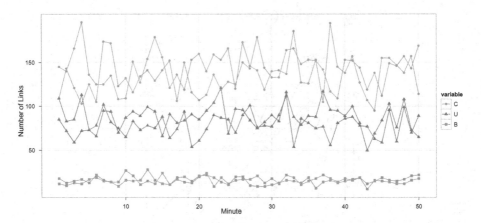

Fig. 14. Measured Links on the lawn for Channels 0 and 3: (C)hanges, (U)nidirectional ones, (B)idirectional ones

between rounds. The figure shows the number of unidirectional (U) and bidirectional (B) links as well as the number of changes between the previous round an the current one (C). Each change of a single link is counted separately, meaning that a unidirectional link that appears or disappears counts as one, a bidirectional one that turns unidirectional is also counted as one but a bidirectional link that appears or disappears counts as two changes, one for each directed link contained therein. It can be seen that the number of link changes is often higher than the number of unidirectional links. This is due to the fact that when one unidirectional links disappears and another appears, two changes occurred.

Figure 15 shows a box plot of the number of changes per round for each environment and channel. It can be seen that apart from the tree environment which only featured 25 nodes instead of 36, the number of changes seems to be fairly independent of the environment.

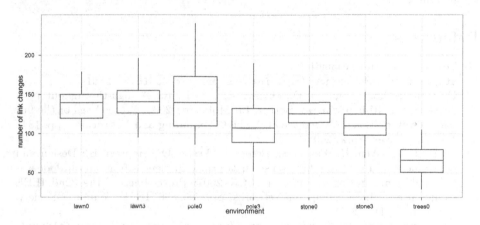

Fig. 15. Link Changes per Round

The ratio of unidirectional links compared to bidirectional ones changes a lot, but there are always far more unidirectional than bidirectional links present. The ratio varies between 3 : 1 and 91 :1, with an average value of 8.69 : 1 over all presented experiments. This high number of unidirectional links makes it necessary to rethink the strategy of ignoring unidirectional links in routing protocols.

When considering the networks consisting of 36 nodes, an average number of 108 link changes per round can be recorded. This high number of link changes in a very short time makes it highly improbable that paths that have been measured at one point in time will exist long enough to transmit a high number of application messages over this exact path. Other forwarding mechanisms are required, which can react to such changes implicitly, without creating route error messages and repetition of route request flooding.

5   Conclusion

In this paper we have presented the results of connectivity measurements we conducted using real sensor network hardware. The experiments were conducted in four different locations and on two different channels. The results show, that unidirectional links are far more common than previously assumed, and link stability does basically not exist. This makes usage of traditional routing protocols in wireless sensor networks hard, to say the least. New protocols need to be devised, that can deal with the influence of an ever changing environment.

Complete connectivity graphs in visual form as well as simulation matrices for OMNeT++ [12] were also generated for all experiments, but are not included here. Complete sets of connectivity data, presented as graphs or connectivity change lists that can be used for simulations, e.g. using OMNeT++, are available upon request.

References

1. Texas instruments cc430f6137,
   http://focus.ti.com/docs/prod/folders/print/cc430f6137.html
2. De Couto, D.S.J., Aguayo, D., Bicket, J., Morris, R.: A high-throughput path metric for multi-hop wireless routing. In: MobiCom 2003: Proceedings of the 9th Annual International Conference on Mobile Computing and Networking, pp. 134–146. ACM, New York (2003)
3. Le Dinh, T., Wen, H., Pavan, S., Peter, C., Leslie, O., Stephen, B.: Design and deployment of a remote robust sensor network: Experiences from an outdoor water quality monitoring network. In: LCN 2007: Proceedings of the 32nd IEEE Conference on Local Computer Networks, pp. 799–806. IEEE Computer Society, Washington, DC (2007)
4. Texas instruments ez430-chronos, http://focus.ti.com/docs/toolsw/folders/print/ez430-chronos.html?DCMP=Chronos&HQS=Other+OT+chronos
5. Johnson, D., Maltz, D., Broch, J.: DSR The Dynamic Source Routing Protocol for Multihop Wireless Ad Hoc Networks, ch.5, pp. 139–172. Addison-Wesley (2001)
6. Marina, M.K., Das, S.R.: Routing performance in the presence of unidirectional links in multihop wireless networks. In: Proceedings of the 3rd ACM International Symposium on Mobile Ad Hoc Networking & Computing, MobiHoc 2002, pp. 12–23. ACM, New York (2002)
7. Perkins, C.E., Royer, E.M.: Ad Hoc on-demand distance vector routing. In: Proceedings of the 2nd IEEE Workshop on Mobile Computing Systems and Applications, New Orleans, LA, pp. 90–100 (February 1999)
8. Sang, L., Arora, A., Zhang, H.: On exploiting asymmetric wireless links via one-way estimation. In: MobiHoc 2007: Proceedings of the 8th ACM International Symposium on Mobile Ad Hoc Networking and Computing, pp. 11–21. ACM Press, New York (2007)
9. Schiller, J., Liers, A., Ritter, H., Winter, R., Voigt, T.: Scatterweb - low power sensor nodes and energy aware routing. In: Proceedings of the 38th Hawaii International Conference on System Sciences (2005)
10. Turau, Renner, Venzke: The heathland experiment: Results and experiences. In: Proceedings of the REALWSN 2005 Workshop on Real-World Wireless Sensor Networks (June 2005)

11. Turau, V.: The heathland experiment: Results and experiences, presentation, http://www.tm.uka.de/forschung/spp1140/events/ kolloquium/2005-11/turau.pdf
12. Varga, A.: The omnet++ discrete event simulation system. In: Proceedings of the European Simulation Multiconference (ESM 2001), Prague, Czech Republic (June 2001)
13. Woo, A., Tong, T., Culler, D.: Taming the underlying challenges of reliable multihop routing in sensor networks. In: SenSys 2003: Proceedings of the 1st International Conference on Embedded Networked Sensor Systems, pp. 14–27. ACM Press, New York (2003)
14. Zhao, J., Govindan, R.: Understanding packet delivery performance in dense wireless sensor networks. In: SenSys 2003: Proceedings of the 1st International Conference on Embedded Networked Sensor Systems, pp. 1–13. ACM Press, New York (2003)

Author Index